SpringerBriefs in Probability and Mathematical Statistics

More information about this series at http://www.springer.com/series/14353

Vladas Pipiras • Murad S. Taqqu

Stable Non-Gaussian Self-Similar Processes with Stationary Increments

Springer

Vladas Pipiras
Statistics and Operations Research
University of North Carolina at Chapel Hill
Chapel Hill, NC, USA

Murad S. Taqqu
Department of Mathematics and Statistics
Boston University
Boston, MA, USA

ISSN 2365-4333 ISSN 2365-4341 (electronic)
SpringerBriefs in Probability and Mathematical Statistics
ISBN 978-3-319-62330-6 ISBN 978-3-319-62331-3 (eBook)
DOI 10.1007/978-3-319-62331-3

Library of Congress Control Number: 2017947156

Printed on acid-free paper

This Springer imprint is published by Springer Nature
The registered company is Springer International Publishing AG
The registered company address is: Gewerbestrasse 11, 6330 Cham, Switzerland

To Terese, Jovita, and Milda
and
to Jeremy and Yael

Preface

Fractional Brownian motion is (up to a constant and for a fixed self-similarity parameter) the unique Gaussian self-similar process with stationary increments. When the assumption of Gaussian distributions is replaced by that of stable (non-Gaussian) distributions, the situation is more complex. This is because there are in fact many different such stable (non-Gaussian) processes (this is for the same self-similarity and stability parameters, discounting multiplicative constants). This work provides a self-contained presentation on the structure of a large class of these stable processes, known as *self-similar mixed moving averages*. These include

- Linear fractional stable motion (LFSM)
- Log-fractional stable motion
- Mixed truncated fractional stable motion
- The Samorodnitsky processes
- The Takenaka processes
- The Telecom process

All these processes are different extensions of fractional Brownian motion to the infinite variance stable case. They are defined through integral representations with respect to a stable non-Gaussian measure. *Minimal integral representations* are introduced first, and then the *rigidity properties of stable processes* are discussed. The rigidity will allow us to take advantage of invariance properties such as self-similarity. We will in fact relate stable processes with an invariance property, such as self-similar mixed moving averages, to nonsingular flows and their functionals. Various decompositions of flows are discussed, including

- Dissipative flows
- Conservative flows
- Periodic flows
- Cyclic flows
- Fixed (identity) flows

The periodic, cyclic, and fixed flows are typical examples of conservative flows. We also provide an example of "the fourth kind," namely a conservative flow which is not one of these. These flows are important because they lead to decompositions of the associated self-similar mixed moving averages in major components.

By using minimal representations and flows, we will be able to show that various processes such as those listed above are different from each other. Minimal representations can thus serve to identify the process but they are not always very easy to determine in practice. We will also provide identification criteria which do not rely on either minimal representations or flows, and which are based instead on the structure of the kernel function in the integral representation of the process.

Chapel Hill, NC, USA Vladas Pipiras
Boston, MA, USA Murad S. Taqqu
August 2017

Acknowledgments

Vladas Pipiras was supported in part by the NSA grant H98230-13-1-0220 and NSF grant DMS-1712966 at the University of North Carolina. Murad S. Taqqu was partially supported by the NSF grants DMS-1007616 and DMS-1309009 at Boston University. The authors would also like to thank Donna Chernyk at Springer for her support and encouragement throughout the process of writing the book.

Contents

Acronyms

FBM	Fractional Brownian motion
CFSM	Conservative fractional stable motion
cLFSM	Cyclic fractional stable motion
(C\P)FSM	Conservative nonperiodic fractional stable motion
DFSM	Dissipative fractional stable motion
FFSM	Fixed fractional stable motion
LFSM	Linear fractional stable motion
PFSM	Periodic fractional stable motion
$S\alpha S$	Symmetric α-stable

Chapter 1
Preliminaries

1.1 Introduction

This book focuses on symmetric α-stable ($S\alpha S$, $\alpha \in (0,2)$), self-similar stochastic processes $\{X_\alpha(t)\}_{t\in\mathbb{R}}$ with stationary increments. Recall that a stochastic process $\{X_\alpha(t)\}_{t\in\mathbb{R}}$ is H-self-similar with the self-similarity parameter $H > 0$ if, for any $c > 0$,

$$\{X_\alpha(ct)\}_{t\in\mathbb{R}} \stackrel{d}{=} \{c^H X_\alpha(t)\}_{t\in\mathbb{R}}, \tag{1.1}$$

where $\stackrel{d}{=}$ denotes the equality of finite-dimensional distributions. See, e.g., Embrechts and Maejima [13] and Pipiras and Taqqu [42]. Stationarity of the increments is defined as follows: for any $h \in \mathbb{R}$,

$$\{X_\alpha(t+h) - X_\alpha(h)\}_{t\in\mathbb{R}} \stackrel{d}{=} \{X_\alpha(t) - X_\alpha(0)\}_{t\in\mathbb{R}}. \tag{1.2}$$

Finally, the process $\{X_\alpha(t)\}_{t\in\mathbb{R}}$ is $S\alpha S$ if its finite-dimensional distributions are $S\alpha S$, that is, their linear combinations are $S\alpha S$ random variables with the characteristic function: for any $\theta, \theta_1, \dots, \theta_n \in \mathbb{R}$, $t_1, \dots, t_n \in \mathbb{R}$, $n \geq 1$,

$$\mathbb{E}\exp\Big\{i\theta \sum_{k=1}^{n} \theta_k X_\alpha(t_k)\Big\} = \exp\Big\{-\sigma^\alpha(\theta_1,\dots,\theta_n,t_1,\dots,t_n)|\theta|^\alpha\Big\}, \tag{1.3}$$

where $\sigma(\theta_1,\dots,\theta_n,\ t_1,\dots,t_n) > 0$ is a scale parameter depending on $t_1,\dots,t_n$, $\theta_1,\dots,\theta_n$. See Section 1.2 below.

The case $\alpha = 2$ of $S\alpha S$ processes corresponds to Gaussian, zero mean processes, but this case will be excluded from this work. In particular, it is well known that in the Gaussian case $\alpha = 2$, the process known as fractional Brownian motion $B_H(t)$, $0 < H < 1$, has mean zero and covariance

$$\mathbb{E}B_H(t)B_H(s) = C\Big(|t|^{2H} + |s|^{2H} - |t-s|^{2H}\Big), \quad s,t \in \mathbb{R},$$

V. Pipiras, M.S. Taqqu, *Stable Non-Gaussian Self-Similar Processes with Stationary Increments*, SpringerBriefs in Probability and Mathematical Statistics,
DOI 10.1007/978-3-319-62331-3_1

where $C > 0$ is a constant. It is (up to a constant and for a fixed self-similarity parameter H) the unique Gaussian self-similar process with stationary increments. See, e.g., Samorodnitsky and Taqqu [56], Embrechts and Maejima [13], Pipiras and Taqqu [42], and Samorodnitsky [54]. The situation for $\alpha \in (0,2)$, which is the focus of this work, is quite different. It turns out that, for a fixed self-similarity parameter H and stability index α, there are many *different* stable self-similar processes with stationary increments. We shall present here a way to describe and classify these processes by relating them to the so-called (deterministic) flows. This approach was initiated by Rosiński [46] for stable stationary processes and applies, more generally, to stable processes possessing an invariance property such as self-similarity or stationarity of increments.

The structure of this book is as follows. We review $S\alpha S$ random variables, stochastic processes, and integrals in Section 1.2, and list in Section 1.3 the basic assumptions which will be valid throughout this work. In Section 2.1, we examine the integral representations of stable processes and introduce representations which are minimal. A stable process has many different integral representations, and its minimal representations play a fundamental role in the sequel. In particular, two minimal representations of a stable process are necessarily related in a very rigid way as described in Section 2.2. It is this rigidity property of minimal representations which will allow us to relate stable processes with an invariance property to flows. Flows and their functionals are discussed in Section 2.3.

In Sections 3.1 and 3.2, we consider a large class of stable self-similar processes with stationary increments which are called *self-similar mixed moving averages*. By using the connection between stable processes and flows, and the structure of these underlying flows, we will be able to identify a number of fundamental classes of the stable processes. Appendix A contains some historical notes on the material of this work. Appendix B gathers some facts on standard Lebesgue spaces and projections that are used in the book. Appendix C summarizes the notation for the subclasses of processes related to various flows.

1.2 $S\alpha S$ Random Variables, Stochastic Processes, and Integral Representations

In this section, we review symmetric α-stable ($S\alpha S$), $\alpha \in (0,2)$, stochastic processes $\{X_\alpha(t)\}_{t\in T}$, and their integral representations. The index set T will usually be taken as the real line $\mathbb{R}$ elsewhere in this work but its structure is irrelevant for this section. In a $S\alpha S$ process $\{X_\alpha(t)\}_{t\in T}$, each random variable $X_\alpha(t)$ is $S\alpha S$, in the sense of the following definition.

Definition 1.1. A random variable X is called *symmetric α-stable* ($S\alpha S$), $\alpha \in (0,2)$, if its characteristic function is given by

$$\mathbb{E}e^{i\theta X} = e^{-\sigma^\alpha |\theta|^\alpha}, \quad \theta \in \mathbb{R}, \tag{1.4}$$

for some $\sigma > 0$, called a *scale* parameter.

Note that if X is $S\alpha S$, then $(-X)$ has the same distribution as X. Such X is called *symmetric*, thus explaining the first part of the term "symmetric α-stable." Note also that, if X is $S\alpha S$, then so is cX for a constant $c \neq 0$. Moreover, if X has a scale parameter σ, then the scale parameter of cX is $|c|\sigma$. This explains why σ is called a scale parameter, that is, scaling X by $c > 0$ has the same effect on its scale parameter. The term "stable" refers to the following property of $S\alpha S$ X that is verified easily through characteristic functions:

$$X + X' \stackrel{d}{=} 2^{1/\alpha} X, \tag{1.5}$$

where X' is an independent copy of X. That is, a $S\alpha S$ distribution is "stable" under the summation in (1.5). The stability property (1.5) can be extended to a finite sum of N independent copies of X, in which case $2^{1/\alpha}$ above needs to be replaced by $N^{1/\alpha}$.

Much is known about $S\alpha S$ random variables. See the monographs by Zolotarev [69], Samorodnitsky and Taqqu [56], and Uchaikin and Zolotarev [63]. For example, the distribution tails of $S\alpha S$ X are characterized by (Property 1.2.15 in Samorodnitsky and Taqqu [56]): for $x > 0$,

$$\frac{1}{2}\mathbb{P}(|X| > x) = \mathbb{P}(X > x) = \mathbb{P}(X < -x) \sim Cx^{-\alpha}, \tag{1.6}$$

as $x \to \infty$, where C is a positive constant and $\sim$ indicates asymptotic equivalence (that is, e.g., $x^{\alpha}\mathbb{P}(X > x) \to C$, as $x \to \infty$). The constant C in (1.6) equals

$$C = \frac{C_\alpha}{2} \quad \text{with} \quad C_\alpha = \begin{cases} \frac{1-\alpha}{\Gamma(2-\alpha)\cos(\pi\alpha/2)}, & \alpha \neq 1, \\ \frac{2}{\pi}, & \alpha = 1, \end{cases} \tag{1.7}$$

with $\Gamma(\cdot)$ denoting the gamma function. The first two equalities in (1.6) hold since X is symmetric. The relation (1.6) yields

$$\mathbb{E}|X|^p = \infty \text{ if } p \in [\alpha, \infty), \quad \mathbb{E}|X|^p < \infty \text{ if } p \in (0, \alpha). \tag{1.8}$$

In particular, $\mathbb{E}X^2 = \infty$ for $\alpha \in (0,2)$, $\mathbb{E}|X| = \infty$ for $\alpha \in (0,1]$ and $\mathbb{E}|X| < \infty$ for $\alpha \in (1,2)$. For $\alpha \in (0,2)$, a $S\alpha S$ distribution has a density, but unless $\alpha = 1/2$ or 1, the density does not have a simple closed form (but see also Górska and Penson [17]). The $S\alpha S$ density with $\alpha = 1$ is the Cauchy density, and with $\alpha = 1/2$, it can be expressed in terms of the Lévy distribution.

$S\alpha S$ random variables can be thought as canonical models for symmetric random variables satisfying (1.6). In fact, they arise in the (non-Gaussian) central limit theorem as possible limits of partials sums, as $n \to \infty$,

$$\frac{1}{b_n}\Big(\sum_{k=1}^{n} X_k - a_n\Big) \stackrel{d}{\to} X, \tag{1.9}$$

where $0 < b_n \to \infty$ and $a_n \in \mathbb{R}$, $\xrightarrow{d}$ indicates the convergence in distribution, and X_k's are i.i.d. random variables. The limiting variable X is $S\alpha S$, $\alpha \in (0,2)$, if and only if, for some constant $c > 0$ and a slowly varying function h,

$$\frac{\mathbb{P}(X_k > x)}{x^{-\alpha}h(x)} \to c, \qquad \frac{\mathbb{P}(X_k < -x)}{x^{-\alpha}h(x)} \to c, \tag{1.10}$$

as $x \to \infty$ (Gnedenko and Kolmogorov [16]). A slowly varying function h is, by definition, positive and satisfies $h(ax)/h(x) \to 1$ for any $a > 0$, as $x \to \infty$; e.g., $h(x) =$ const or $h(x) = \ln x$, $x > 1$. Moreover, when $\alpha \neq 1$, the sequences a_n and b_n in (1.9) can be taken as

$$nh(b_n) = b_n^{\alpha}, \quad a_n = \begin{cases} 0, & 0 < \alpha < 1, \\ \mathbb{E}X_k, & 1 < \alpha < 2. \end{cases} \tag{1.11}$$

When $\alpha = 1$, the sequence b_n can be chosen as in (1.11) but the situation with a_n is more involved (Aaronson and Denker [2]). One can show that b_n can be written as

$$b_n = n^{1/\alpha} h^*(n)^{1/\alpha}, \tag{1.12}$$

for another slowly varying function h^* (see, e.g., Appendix A.4 in Pipiras and Taqqu [42]).

From another important perspective, $S\alpha S$ random variables can be viewed as special cases of infinitely divisible random variables. See, e.g., the monograph by Sato [57] for more information on infinitely divisible distributions. In particular, the so-called Lévy-Khintchine representation of an infinitely divisible random variable X states that the characteristic function of X can be expressed as

$$\mathbb{E}e^{i\theta X} = \exp\left\{ i\gamma\theta + \frac{\delta^2\theta^2}{2} + \int_{\mathbb{R}\setminus\{0\}} (e^{i\theta x} - 1 - i\theta x 1_{|x|\le 1})\nu(dx) \right\}, \quad \theta \in \mathbb{R}, \tag{1.13}$$

where $\gamma \in \mathbb{R}$ is a location parameter, $\delta > 0$ is a parameter associated with the Gaussian "part" of X, and $\nu(dx)$, called the Lévy measure, satisfies $\int_{\mathbb{R}} \min\{1, |x^2|\}\nu(dx) < \infty$. A $S\alpha S$ X is infinitely divisible with

$$\gamma = 0, \quad \delta = 0, \quad \nu(dx) = \frac{c_0 dx}{|x|^{\alpha+1}} \tag{1.14}$$

(see Sato [57], Section 14). Yet another interesting series expansion of $S\alpha S$ random variables is due to Raoul LePage; see Sections 1.4–1.5 in Samorodnitsky and Taqqu [56].

Remark 1.1. $S\alpha S$ random variables are important special cases of more general, not necessarily symmetric, α-stable random variables. An α-stable random variable is characterized similarly by a scale parameter $\sigma > 0$ and also by a location parameter $\gamma \in \mathbb{R}$ and a skewness parameter $\beta \in [-1, 1]$. See Zolotarev [69], Samorodnitsky and Taqqu [56], or Uchaikin and Zolotarev [63] for a general definition. While γ just shifts the distribution on the real axis, the parameter β controls its asymmetry. When $\gamma = 0$ and $\beta = 0$, an α-stable random variable is $S\alpha S$. We focus on the $S\alpha S$

case for simplicity. Some results for general α-stable processes in line with the approach described in this work can be found in Rosiński [45].

We now consider not just one random variable but a collection of random variables, also known as a stochastic process. Let T be an index set, e.g., $T = \mathbb{R}$.

Definition 1.2. A stochastic process $\{X_\alpha(t)\}_{t\in T}$ is called *symmetric α-stable* ($S\alpha S$), $\alpha \in (0,2)$, if every linear combination $\sum_{k=1}^n \theta_k X_\alpha(t_k)$, $t_1,\ldots,t_n \in T$, $\theta_1,\ldots,\theta_n \in \mathbb{R}$, is a $S\alpha S$ random variable. That is,

$$\mathbb{E}\exp\Big\{i\theta\sum_{k=1}^n \theta_k X_\alpha(t_k)\Big\} = \exp\Big\{-\sigma^\alpha(\theta_1,\ldots,\theta_n,t_1,\ldots,t_n)|\theta|^\alpha\Big\},\quad \theta\in\mathbb{R}, \tag{1.15}$$

where $\sigma(\theta_1,\ldots,\theta_n,t_1,\ldots,t_n) > 0$ is a scale parameter.

Remark 1.2. Nonsymmetric α-stable processes cannot be always defined using conditions involving linear combinations (see Samorodnitsky and Taqqu [56]), as is the case for Gaussian processes.

Example 1.1. A random bivariate vector $(X_1,X_2)'$[1] is $S\alpha S$ if, for all $\theta_1,\theta_2\in\mathbb{R}$,

$$\mathbb{E}\exp\Big\{i\theta(\theta_1X_1+\theta_2X_2)\Big\} = \exp\Big\{-\sigma^\alpha(\theta_1,\theta_2)|\theta|^\alpha\Big\},\quad \theta\in\mathbb{R}.$$

For example, let ξ_j's be i.i.d. $S\alpha S$ random variables with scale parameter 1 and set

$$X_1=\sum_j f_1(j)\xi_j,\quad X_2=\sum_j f_2(j)\xi_j, \tag{1.16}$$

where $f_k(j)\in\mathbb{R}$ satisfy $\sum_j |f_k(j)|^\alpha<\infty$, $k=1,2$. The latter conditions ensure that X_1 and X_2 are well defined, in the case when the sum $\sum_j$ is countable. Then, $(X_1,X_2)'$ is $S\alpha S$ since, by independence of the ξ_j's,

$$\mathbb{E}\exp\Big\{i\theta(\theta_1X_1+\theta_2X_2)\Big\} = \mathbb{E}\exp\Big\{i\theta\sum_j(\theta_1f_1(j)+\theta_2f_2(j))\xi_j\Big\}$$

$$=\exp\Big\{-|\theta|^\alpha\sum_j|\theta_1f_1(j)+\theta_2f_2(j)|^\alpha\Big\} = \exp\Big\{-\sigma^\alpha(\theta_1,\theta_2)|\theta|^\alpha\Big\},$$

where

$$\sigma(\theta_1,\theta_2)=\Big(\sum_j|\theta_1f_1(j)+\theta_2f_2(j)|^\alpha\Big)^{1/\alpha}.$$

It turns out that any $S\alpha S$ process can essentially be represented in a suitable extension of (1.16), where the sum is possibly replaced by an integral and the i.i.d. $S\alpha S$ variables ξ_j's by a $S\alpha S$ random measure. Such a measure is defined next, followed by the definition of an integral. Let $(S,\mathscr{S},m)$ be a measure space, and

$$\mathscr{S}_0=\{A\in\mathscr{S}: m(A)<\infty\}.$$

[1] Prime denotes the transpose.

Definition 1.3. A *symmetric α-stable* ($S\alpha S$), $\alpha \in (0,2)$, random measure M on the measurable space $(S,\mathscr{S})$ with control measure m is a mapping

$$M : \mathscr{S}_0 \to L^0(\Omega),$$

where $L^0(\Omega)$ represents the collection of random variables, such that

(i) it is σ-additive, that is,

$$M(\cup_i A_i) = \sum_i M(A_i) \quad \text{a.s.}$$

for all disjoint $A_i \in \mathscr{S}_0$ such that $\cup_i A_i \in \mathscr{S}_0$;

(ii) it is independently scattered, that is, the random variables $M(A_1),\dots,M(A_n)$ are independent for all disjoint $A_i \in \mathscr{S}_0$, $i = 1,\dots,n$;

(iii) for any $A \in \mathscr{S}_0$, $M(A)$ is a $S\alpha S$ random variable with a scale parameter $m(A)$.

Informally, one thinks of M as a "random (signed) measure" $M(ds)$ such that $M(ds)$ and $M(ds')$ are independent $S\alpha S$ random variables with scale parameters $m(ds)$ and $m(ds')$, respectively, as long as ds and ds' are disjoint. Existence of $S\alpha S$ random measures is discussed in Samorodnitsky and Taqqu [56].

Example 1.2. Let $S = \mathbb{Z}$, $\mathscr{S} = 2^{\mathbb{Z}}$ (the collection of all possible subsets of $\mathbb{Z}$) and $m(ds) =$ counting measure (i.e., $m(\{j\}) = 1$ for all $j \in \mathbb{Z}$). A $S\alpha S$ random measure M on $(S,\mathscr{S})$ with control measure m can be thought as consisting of i.i.d. $S\alpha S$ random variables $\xi_j = M(\{j\})$, $j \in \mathbb{Z}$, with scale parameter $m(\{j\}) = 1$. Then, the random vector $(X_1,X_2)'$ in Example 1.1 can also be expressed as

$$X_t = \sum_j f_t(j)M(\{j\}), \quad t = 1,2. \tag{1.17}$$

A $S\alpha S$ random measure M on $(S,\mathscr{S})$ with control measure m can be used to define an integral

$$I(f) = \int_S f(s)M(ds), \tag{1.18}$$

where $f : S \to \mathbb{R}$ is a deterministic function such that

$$f \in L^\alpha(S,\mathscr{S},m), \quad \text{that is,} \quad \int_S |f(s)|^\alpha m(ds) < \infty.$$

One way this could be done is by defining first the integral $I(f)$ for "simple" functions

$$f(s) = \sum_{k=1}^n f_k 1_{A_k}(s), \quad f_k \in \mathbb{R},\ A_k \in \mathscr{S}_0,$$

in a natural way by setting

$$I(f) = \sum_{k=1}^n f_k M(A_k), \tag{1.19}$$

where 1_A denotes the indicator function of a set A. Such $I(f)$ is $S\alpha S$ with scale parameter $\sum_{k=1}^n |f_k|^\alpha m(A_k) = \int_S |f(s)|^\alpha m(ds)$. Then, a limiting argument is used to extend the integral to all $f \in L^\alpha(S,\mathscr{S},m)$; see Samorodnitsky and Taqqu [56] for details. The integral is constructed so that

$$I(af+g) = aI(f)+I(g) \quad \text{a.s.} \tag{1.20}$$

where $a \in \mathbb{R}$ and $f,g \in L^\alpha(S,\mathscr{S},m)$.

Definition 1.4. If M is a $S\alpha S$ random measure on $(S,\mathscr{S})$ with control measure m, the integral $I(f)$ in (1.18) is defined for $f \in L^\alpha(S,\mathscr{S},m)$ as a $S\alpha S$ random variable with the scale parameter $\int_S |f(s)|^\alpha m(ds)$. Moreover, the integral has the properties (1.19) and (1.20).

Informally, as with any integral, one thinks of (1.18) as

$$\sum_k f(s_k)M(ds_k),$$

where disjoint ds_k's form a partition of S, and the point s_k falls into ds_k.

Example 1.3. The representation (1.17) of the vector $(X_1,X_2)'$ in Example 1.2 can also be expressed as

$$X_t = \int_{\mathbb{Z}} f_t(s)M(ds), \quad t = 1,2, \tag{1.21}$$

where M is a $S\alpha S$ random measure on $\mathbb{Z}$ with a counting control measure.

Stochastic integrals present a convenient way to define and to work with $S\alpha S$ stochastic processes. Note, in particular, that taking a collection of deterministic functions $\{f_t\}_{t\in T} \subset L^\alpha(S,\mathscr{S},m)$, that is, $\int_S |f_t(s)|^\alpha m(ds) < \infty$ for all t, and a $S\alpha S$ random measure M on $(S,\mathscr{S})$ with control measure m, one can define a stochastic process

$$X_\alpha(t) = \int_S f_t(s)M(ds), \quad t \in T. \tag{1.22}$$

The process $\{X_\alpha(t)\}_{t\in T}$ is $S\alpha S$ since any linear combination $\sum_{k=1}^n \theta_k X_\alpha(t_k)$ is $S\alpha S$, which follows from

$$\mathbb{E}\exp\Big\{i\theta\sum_{k=1}^n \theta_k X_\alpha(t_k)\Big\} = \mathbb{E}\exp\Big\{i\theta\int_S\sum_{k=1}^n \theta_k f_{t_k}(s)M(ds)\Big\}$$

$$= \exp\Big\{-|\theta|^\alpha\int_S\Big|\sum_{k=1}^n \theta_k f_{t_k}(s)\Big|^\alpha m(ds)\Big\} = \exp\Big\{-\Big\|\sum_{k=1}^n \theta_k f_{t_k}\Big\|_\alpha^\alpha |\theta|^\alpha\Big\},$$

where

$$\|g\|_\alpha^\alpha = \int_S |g(s)|^\alpha m(ds).$$

If the underlying space is to be emphasized, we shall also write $\|g\|_\alpha^\alpha = \|g\|_{L^\alpha(S,m)}^\alpha$. The random variables $X_\alpha(t)$'s are dependent in general but the representation (1.22) provides a convenient way to "decorrelate" them into independently scattered random measure $M(ds)$. Any dependence of the random variables $X_\alpha(t)$'s is now captured by the deterministic functions $f_t(s)$. In fact, as mentioned in Section 1.3 below, it is known that under mild technical assumptions, any $S\alpha S$ process can be represented in distribution as in (1.22). A comprehensive reference on stable processes and their integral representations is Samorodnitsky and Taqqu [56].

1.3 Basic Assumptions

As noted in Sections 1.1 and 1.2, we shall focus throughout this book on symmetric α-stable ($S\alpha S$), $\alpha \in (0,2)$, processes. The fact that the Gaussian case $\alpha = 2$ is not included is critical because many of the results we will obtain will not hold for Gaussian processes. The processes will be denoted $X_\alpha = \{X_\alpha(t)\}_{t\in T}$, where T is an index set, typically $T = \mathbb{R}$. It will be assumed that T is a separable[2] metric space, equipped with a σ-field $\mathscr{B}(T) = \sigma\{A : A \subset T \text{ is open}\}$. In the case of $T = \mathbb{R}$ considered below, the metric is the usual Euclidean metric.

We shall work with $S\alpha S$ processes through their integral representations as

$$\{X_\alpha(t)\}_{t\in T} \stackrel{d}{=} \left\{ \int_S f_t(s) M(ds) \right\}_{t\in T}, \tag{1.23}$$

where M is a $S\alpha S$ random measure on $(S, \mathscr{S})$ with control measure m, and $\{f_t\}_{t\in T} \subset L^\alpha(S,m)$ (see Section 1.2 above). The relation (1.23) is equivalent to the characteristic function of a linear combination $\sum_{k=1}^n \theta_k X_\alpha(t_k)$, $\theta_k \in \mathbb{R}$, $t_k \in T$, having the form

$$\mathbb{E}\exp\left\{ i \sum_{k=1}^n \theta_k X_\alpha(t_k) \right\} = \exp\left\{ -\int_S \left| \sum_{k=1}^n \theta_k f_{t_k}(s) \right|^\alpha m(ds) \right\} = \exp\left\{ -\left\| \sum_{k=1}^n \theta_k f_{t_k} \right\|_\alpha^\alpha \right\}. \tag{1.24}$$

For technical reasons, we shall assume that $(S, \mathscr{S}, m)$ is a *standard Lebesgue space*, where S is a Borel subset of a Polish space,[3] the σ-field $\mathscr{S}$ is the σ-field of Borel sets $\mathscr{B}(S)$ defined as $\mathscr{B}(S) = \sigma\{A : A \subset S \text{ is open}\}$, and m is a σ-finite measure. For example, S could be the Euclidean space $\mathbb{R}^n$ with a measure consisting of a mixture of the Lebesgue measure and discrete point masses. See Appendix B for some basic information about standard Lebesgue spaces. Unless indicated otherwise, measurability will be understood throughout to be with respect to Borel σ-fields. In particular, we assume that the map $f_t(s) : T \times S \to \mathbb{R}$ is $\mathscr{B}(T \times S)$-measurable. When measurability is meant with respect to the completion σ-field

[2] The space is *separable* if it contains a countable, dense set.

[3] A metric space is Polish if it is separable and complete. The space is complete if all Cauchy sequences in the space converge to an element of the space.

under a measure m, we will write "m-measurable."[4] To avoid unnecessary details, we also suppose without loss of generality that

$$\mathrm{supp}\{f_t, t \in T\} = S \quad m\text{-a.e.}, \tag{1.25}$$

where $\mathrm{supp}\{f_t, t \in T\}$, the support of f_t, $t \in T$, is a minimal (m-a.e.) set $A \in \mathscr{B}(S)$ such that

$$m\{f_t(s) \neq 0, s \notin A\} = 0 \quad \text{for every } t \in T.$$

Informally, this last relation states that outside the set A, the values of the function f_t must be 0 for every t.

It is known that every measurable, real-valued, $S\alpha S$ process X_α has an integral representation (1.23) with, for example, the standard Lebesgue space $S = (0,1)$, $\mathscr{S} = \mathscr{B}(0,1)$, $m =$ Lebesgue measure and a measurable $f_t(s)$ (see Samorodnitsky and Taqqu [56], Theorems 13.2.1 and 9.4.2). On the other hand, most examples of stable processes are also defined through integral representations of the type (1.23) on a standard Lebesgue space and measurable $f_t(s)$.

The following lemma will be used many times.

Lemma 1.1. *Suppose that $(S_1, \mathscr{S}_1, m_1)$ and $(S_2, \mathscr{S}_2, m_2)$ are two measure spaces. Let $F : S_1 \times S_2 \to \mathbb{R}$ be a measurable function on $(S_1 \times S_2, \mathscr{S}_1 \times \mathscr{S}_2, m_1 \times m_2)$. Then,*

(i) if for almost every $s_1 \in S_1$, $F(s_1,s_2) = 0$ a.e. $m_2(ds_2)$, we have $F(s_1,s_2) = 0$ a.e. $(m_1 \times m_2)(ds_1,ds_2)$, and, conversely,
(ii) if $F(s_1,s_2) = 0$ a.e. $(m_1 \times m_2)(ds_1,ds_2)$, then for almost every $s_1 \in S_1$, $F(s_1,s_2) = 0$ a.e. $m_2(ds_2)$.

PROOF: We first prove (i). Set $A = \{(s_1,s_2) : F(s_1,s_2) \neq 0\}$ and note that by the assumption, for almost every $s_1 \in S_1$, $\int_{S_2} 1_A(s_1,s_2) m_2(ds_2) = 0$. Then, by using Fubini's theorem,

$$\int_{S_1 \times S_2} 1_A(s_1,s_2)\,(m_1 \times m_2)(ds_1,ds_2) = \int_{S_1} \left\{ \int_{S_2} 1_A(s_1,s_2) m_2(ds_2) \right\} m_1(ds_1) = 0.$$

Since the integrand above is positive, we get $1_A(s_1,s_2) = 0$ a.e. $(m_1 \times m_2)(ds_1,ds_2)$. This yields (i). The proof of (ii) is similar. □

[4] To complete a σ-field $\mathscr{F}$ under a measure m include all the sets $A \subset B$ where $B \in \mathscr{F}$ and $m(B) = 0$. Sets in this augmented σ-field are said to be m-measurable.

Chapter 2
Minimality, Rigidity, and Flows

2.1 Minimal Integral Representations

A $S\alpha S$ process X has many integral representations $\{f_t\}_{t\in T}$ on possibly different spaces $(S,\mathscr{B}(S),m)$. For example, another representation can be obtained from (1.23) by mapping $(S,\mathscr{B}(S),m)$ to a different space (that is, by simply making a change of variables). More specifically, if $(\widetilde{S},\mathscr{B}(\widetilde{S}),\widetilde{m})$ is another standard Lebesgue space and $\Phi:\widetilde{S}\to S$ is a one-to-one, onto map with a measurable inverse such that $m\circ\Phi \ll \widetilde{m}$,[1] then

$$\int_S \Big|\sum_{k=1}^n \theta_k f_{t_k}(s)\Big|^\alpha m(ds) = \int_{\widetilde{S}} \Big|\sum_{k=1}^n \theta_k f_{t_k}(\Phi(\widetilde{s}))\Big|^\alpha \frac{d(m\circ\Phi)}{d\widetilde{m}}(\widetilde{s})\widetilde{m}(d\widetilde{s}) \tag{2.1}$$

and hence

$$\Big\{\int_S f_t(s)M(ds)\Big\}_{t\in T} \stackrel{d}{=} \Big\{\int_{\widetilde{S}} f_t(\Phi(\widetilde{s}))\Big\{\frac{d(m\circ\Phi)}{d\widetilde{m}}(\widetilde{s})\Big\}^{1/\alpha}\widetilde{M}(d\widetilde{s})\Big\}_{t\in T}, \tag{2.2}$$

where $\widetilde{M}$ is a $S\alpha S$ random measure on $\widetilde{S}$ with control measure $\widetilde{m}$. The ratio

$$\frac{d(m\circ\Phi)}{d\widetilde{m}}$$

in (2.1) plays essentially the role of a Jacobian of the transformation, allowing us to go from an old space $\widetilde{S}$ to a new space S. This is reflected in (2.2) by the same term but with an $1/\alpha$ power.

Among all integral representations, the so-called *minimal* integral representations play a fundamental role. Minimal representations were introduced by Hardin [20], and subsequently studied in depth by Rosiński [48]. As shown in Hardin [20], every

[1]A measure m_1 is absolutely continuous with respect to a measure m_2, written $m_1 \ll m_2$, if for every measurable set A, $m_2(A)=0$ implies $m_1(A)=0$.

V. Pipiras, M.S. Taqqu, *Stable Non-Gaussian Self-Similar Processes with Stationary Increments*, SpringerBriefs in Probability and Mathematical Statistics,
DOI 10.1007/978-3-319-62331-3_2

separable in probability[2] $S\alpha S$ process has a minimal integral representation. Two, commonly used, equivalent ways to define minimal representations are as follows. A proof of the equivalence is given in Proposition 2.1 below. Another characterization of minimal representations will be given in Section 2.2 below.

Definition 2.1. An integral representation $\{f_t\}_{t\in T}$ in (1.23) is *minimal* if and only if

$$\sigma\Big\{\frac{f_u}{f_v},\ u,v\in T\Big\} = \mathscr{B}(S) \mod m \tag{2.3}$$

or, if and only if, for every nonsingular map $\phi : S \to S$ and $h : S \to \mathbb{R}\setminus\{0\}$ such that, for each $t \in T$, the relation

$$f_t(s) = h(s) f_t(\phi(s)) \quad \text{a.e. } m(ds) \tag{2.4}$$

holds, one has

$$\phi(s) = s \quad \text{a.e. } m(ds). \tag{2.5}$$

(As part of the minimality assumption, it is also assumed implicitly that the condition (1.25) holds.)

The definition (2.3) states that the σ-field $\mathscr{B}(S)$ is generated by ratios of the form f_u/f_v of the family of functions f_t, $t \in T$. The definition (2.4)–(2.5) will be illustrated in the examples below. If (2.4) holds, then

$$\frac{f_{t_1}}{f_{t_2}}(s) = \frac{f_{t_1}}{f_{t_2}}(\phi(s)) \quad \text{a.e. } m(ds) \quad \text{for any } t_1,t_2 \in T. \tag{2.6}$$

The definition states that the only way this can happen under minimality is with ϕ being the identity map.

We write $\mathscr{A} = \mathscr{B} \mod m$ for two σ-fields $\mathscr{A}$ and $\mathscr{B}$ when, for $A \in \mathscr{A}$, there is $B \in \mathscr{B}$ such that $m(A\triangle B) = 0$. Thus informally, from the perspective of the measure m, the sets A and B are essentially the same. If $f_v(s) = 0$, then it is assumed that $(f_u/f_v)(s) = \partial$, the infinity point of the one-point compactification of $\mathbb{R}$. A map $\phi : S \to S$ is *nonsingular* when $m(A) = 0$ implies $m(\phi^{-1}(A)) = 0$. In other words, it cannot happen that sets of positive measure are mapped by ϕ to sets of null measure. We shall also express (2.5) as $\phi = Id$ m-a.e. ("Id" for "identity") and express its opposite as $\phi \neq Id$ m-a.e.

Example 2.1. Definition 2.1 can be understood easily and intuitively through the following example. Consider the $S\alpha S$ processes

$$X_t = \sum_{n\in\mathbb{Z}} a_{t,n}\varepsilon_n = \int_{\mathbb{Z}} f_t(n) M(dn), \quad t \in \mathbb{Z}, \tag{2.7}$$

[2]A process $\{X(t)\}_{t\in T}$ is separable in probability if there is a countable subset $T_0 \subset T$ such that for every $t \in T$, $X(t)$ is the limit in probability of $X(t_k)$ for some $t_k \in T$. By Remark 2 in Samorodnitsky and Taqqu [56], p. 153, separability in probability is equivalent to the so-called S condition. The minimality result of Hardin [20], in fact, assumes that the S condition holds.

where ε_n are independent, $S\alpha S$ random variables with scale parameter 1, M has the control measure $m(dn) = \delta_{\mathbb{Z}}(dn)$ and $f_t(n) = a_{t,n}$. The measure m is discrete and gives a value 1 to points in $\mathbb{Z}$. If (2.7) is nonminimal, then there is $\phi : \mathbb{Z} \to \mathbb{Z}$ such that $\phi \neq Id$ and (2.4) holds. In particular, there is $n_0 \in \mathbb{Z}$ with $m_0 = \phi(n_0) \neq n_0$ such that

$$a_{t,n_0} = c\, a_{t,m_0} \quad (c \neq 0), \tag{2.8}$$

for all t, that is, the coefficient a_{t,m_0} is proportional to another coefficient a_{t,n_0} for all t.[3] One can thus omit the term $a_{t,m_0}\varepsilon_{m_0}$ from the representation (2.7) by adjusting the term $a_{t,n_0}\varepsilon_{n_0}$ accordingly. Since (2.8) holds,

$$a_{t,n_0}\varepsilon_{n_0} + a_{t,m_0}\varepsilon_{m_0} = a_{t,n_0}\varepsilon_{n_0} + c^{-1}a_{t,n_0}\varepsilon_{m_0} \stackrel{d}{=} (1+|c|^{-\alpha})^{1/\alpha} a_{t,n_0}\varepsilon_{n_0},$$

where $\stackrel{d}{=}$ denotes the equality of the finite-dimensional distributions. The representation (2.7) can thus be reduced to

$$X_t \stackrel{d}{=} \sum_{n \neq m_0} \widetilde{a}_{t,n}\varepsilon_n, \tag{2.9}$$

where

$$\widetilde{a}_{t,n} = \begin{cases} a_{t,n}, & \text{if } n \neq n_0, \\ (1+|c|^{-\alpha})^{1/\alpha} a_{t,n}, & \text{if } n = n_0. \end{cases} \tag{2.10}$$

In view of (2.8)–(2.10), nonminimality can thus be viewed as a type of redundancy of integral representations.

This example also sheds some light on the relationship between the conditions (2.3) and (2.4)–(2.5). If (2.4)–(2.5) is not valid, then the relation (2.8) holds. In particular,

$$\frac{f_u}{f_v}(n_0) = \frac{f_u}{f_v}(m_0), \quad \text{all } u, v \in \mathbb{Z}.$$

This implies that the σ-field $\sigma\{f_u/f_v,\ u,v \in \mathbb{Z}\}$ is not fine enough to include the sets $\{n_0\}$ and $\{m_0\}$. Hence, the relation (2.3) does not hold either since $\mathscr{B}(\mathbb{Z})$ consists of all the subsets of $\mathbb{Z}$, including the sets $\{n_0\}$ and $\{m_0\}$.

Example 2.2. Consider a $S\alpha S$ (stationary) moving average process given by

$$X_\alpha(t) = \int_{\mathbb{R}} f(t-u)M(du), \tag{2.11}$$

where $f \in L^\alpha(\mathbb{R}, du)$ and M is a $S\alpha S$ random measure on $\mathbb{R}$ with the Lebesgue control measure du. The integral representation (2.11) is always minimal (excluding the degenerate case $f = 0$ a.e.). Indeed, as in (2.4), suppose by contradiction that there are $\phi : \mathbb{R} \to \mathbb{R}$ and $h : \mathbb{R} \to \mathbb{R} \setminus \{0\}$ such that, for all $t \in \mathbb{R}$,

$$f(t-u) = h(u) f(t - \phi(u)) \quad \text{a.e. } du, \tag{2.12}$$

[3] The fact that (2.8) holds for all t is critical.

with $\phi \neq Id$ a.e. By Lemma 1.1, (i), the relation (2.12) also holds a.e. $dtdu$.[4] By making the change of variables $t-u=v$ and $u=u$, we then have

$$f(v)=h(u)f(v+u-\phi(u)) \quad \text{a.e. } dudv. \tag{2.13}$$

Since $\phi \neq Id$ a.e., there is u_0 such that $\delta_0 = u_0-\phi(u_0) \neq 0$ and (2.13) holds a.e. dv. But then

$$f(v)=h(u_0)f(v-\delta_0) \text{ a.e. } dv,$$

and therefore

$$\int_{\mathbb{R}} |f(v)|^\alpha dv = \sum_{n=-\infty}^{\infty} |h(u_0)|^\alpha \int_{(n-1)\delta_0}^{n\delta_0} |f(v-\delta_0)|^\alpha dv$$

$$= \sum_{n=-\infty}^{\infty} |h(u_0)|^\alpha \int_{(n-2)\delta_0}^{(n-1)\delta_0} |f(v)|^\alpha dv = \sum_{n=-\infty}^{\infty} |h(u_0)|^{2\alpha} \int_{(n-2)\delta_0}^{(n-1)\delta_0} |f(v-\delta_0)|^\alpha dv$$

$$= \ldots = \int_0^{\delta_0} |f(v)|^\alpha dv \sum_{n=-\infty}^{\infty} |h(u_0)|^{\alpha n} = \infty, \tag{2.14}$$

which contradicts the assumption $f \in L^\alpha(\mathbb{R}, du)$. Finally, to verify that the condition (1.25) holds,[5] let

$$A = \operatorname{supp}\{f(t-\cdot), t \in \mathbb{R}\}.$$

Then, for each $t \in \mathbb{R}$, $1_A(u) = 1_A(u+t)$ a.e. du. By Lemma 1.1, (i), $1_A(u) = 1_A(u+t)$ a.e. $dudt$ and, by making the change of variables, $1_A(u) = 1_A(v)$ a.e. $dudv$. Now, fix $v = v_0 \in A$ such that $1_A(u) = 1_A(v_0)$ a.e. du (which is possible since $f \neq 0$ a.e. and hence A has positive Lebesgue measure). Since $v_0 \in A$, we have $1_A(u) = 1$ a.e. du. This yields $A = \mathbb{R}$ a.e., proving (1.25).

Example 2.3. The $S\alpha S$ Lévy motion can be defined as the process having integral representation

$$\int_0^t M(du), \quad t>0,$$

where $M(du)$ is a $S\alpha S$ random measure on $\mathbb{R}$ with the Lebesgue control measure du. It can be seen from definition (see (2.3) and (2.4)–(2.5)) that this integral representation of the $S\alpha S$ Lévy motion is minimal.[6]

---

[4]Indeed, if $F(t,u) = f(t-u) - h(u)f(t-\phi(u))$, then for all $t \in \mathbb{R}$, $F(t,u) = 0$ a.e. du. By Lemma 1.1, (i), we have $F(t,u) = 0$ a.e. $dtdu$.

[5]Recall that this is part of Definition 2.1.

[6]Note also that minimality follows directly from Example 3.3 since the stable Lévy motion is stationary increments moving average.

Proposition 2.1. *The conditions (2.3) and (2.4)–(2.5) are equivalent.*

PROOF: The equivalence of the conditions is proved in Rosiński [48], Proposition 5.1. We include here a proof that (2.3) implies (2.4)–(2.5). Though the idea for proving the converse is straightforward,[7] the actual proof is fairly technical (see the necessity part of Proposition 5.1, Rosiński [48]).

Arguing by contradiction, suppose there is nonsingular ϕ such that (2.4) holds and $\phi \neq Id$ m-a.e. Let K_n, $n \geq 1$, be a sequence of sets separating points[8] of S and generating $\mathscr{B}(S)$. Note that

$$\{s : \phi(s) \neq s\} \subset \bigcup_{n\geq 1} K_n \cap \phi^{-1} K_n^c,$$

where K_n^c denotes the complement of K_n. (Indeed, if s is such that $s \neq \phi(s)$, there is K_n such that $s \in K_n$ and $\phi(s) \notin K_n$ or, equivalently, $s \in K_n$ and $s \in \phi^{-1}K_n^c$.) Since $m\{s : \phi(s) \neq s\} > 0$ by assumption, it follows that

$$m(K_n \cap \phi^{-1} K_n^c) > 0$$

for some n. Now, set $A = K_n \cap \phi^{-1} K_n^c$. Since $A \in \mathscr{B}(S)$, it is enough to show that

$$A \notin \sigma\{f_u/f_v,\ u,v \in T\} \quad m\text{-a.e.}, \tag{2.15}$$

since this contradicts (2.3).

Set $\mathbb{R}_* = \mathbb{R} \cup \{\partial\}$ as explained following Definition 2.1. Recall that f_u and f_v map S into $\mathbb{R}$. A general set of $\sigma\{f_u/f_v,\ u,v \in T\}$ can then be expressed as $F^{-1}B$, where $B \in \mathscr{B}(\mathbb{R}_*^\infty)$ and

$$F = \left(\frac{f_{u_1}}{f_{v_1}}, \frac{f_{u_2}}{f_{v_2}}, \dots\right).$$

Both A and $F^{-1}(B)$ are sets in S. The relation (2.15) is then equivalent to

$$m(A \triangle F^{-1}B) > 0 \tag{2.16}$$

for any F and B as above.

Arguing by contradiction, suppose $m(A \triangle F^{-1}B) = 0$ for some F and B. Then,

$$m(A^c \cap F^{-1}B) = 0 \quad \text{and} \quad m(A \cap F^{-1}B^c) = 0.$$

---

[7] One argues first that it is enough to consider the relation (2.3) for a countable number of pairs $(u_1,v_1),(u_2,v_2),\dots$ and the relation (2.4) for a countable number of points $t = u_1,v_1,u_2,v_2,\dots$. Setting $F = (f_{u_1}/f_{v_1}, f_{u_2}/f_{v_2},\dots)$, which is now a map with values in $Y := (\mathbb{R} \cup \{\partial\})^\infty$, the relation (2.3) becomes $F^{-1}(\mathscr{B}(Y)) = \mathscr{B}(S)$ mod m (that is, for every $A \in F^{-1}(\mathscr{B}(Y))$, there is $B \in \mathscr{B}(S)$ so that $m(A \triangle B) = 0$), and (2.4) can be written as $F(\phi(s)) = F(s)$ a.e. m. Then, arguing by contradiction and quoting Rosiński [48] in slightly adapted form, the idea of the proof is: "Since $F^{-1}(\mathscr{B}(Y)) \neq \mathscr{B}(S)$, the partition $\{F^{-1}\{y\} : y \in Y\}$ contains sufficiently many sets consisting of more than just one point. On each such set one can define an isomorphism different from the identity. Then a function ϕ is obtained by pasting together such isomorphisms."

[8] That is, for any $s_1 \neq s_2$ in S, there is a set K_n such that $s \in K_n$ and $s_2 \notin K_n$ for some n.

Recall that $A = K_n \cap \phi^{-1}K_n^c$. If $s \in A$, then $\phi(s) \in K_n^c$, which implies $\phi(s) \in A^c$. This yields $A \subset \phi^{-1}A^c$. By using (2.4), we can go from the map F to the map $F \circ \phi$ and vice versa. Hence $F = F \circ \phi$ m-a.e. Hence, $F^{-1}B \subset \phi^{-1}(F^{-1}B)$ m-a.e. By intersecting the last two inclusions, we obtain that

$$A \cap F^{-1}B \subset \phi^{-1}(A^c \cap F^{-1}B).$$

It follows that

$$m(\phi^{-1}(A^c \cap F^{-1}B)) \geq m(A \cap F^{-1}B) = m(A \cap F^{-1}B) + m(A \cap F^{-1}B^c) = m(A) > 0,$$

where we used $m(A \cap F^{-1}B^c) = 0$. Thus, by the nonsingularity of ϕ, $m(A^c \cap F^{-1}B) > 0$, which contradicts $m(A \triangle F^{-1}B) = 0$. □

Remark 2.1. Proposition 5.1 of Rosiński [48] referenced above, in fact, leads to a weaker equivalent condition of minimal representations: the representation $\{f_t\}_{t\in T}$ (with full support) is minimal if and only if the relation (2.4) with a null-isomorphism (mod 0) ϕ satisfying

$$\phi \circ \phi = Id \tag{2.17}$$

entails $\phi = Id$ m-a.e.[9] The condition (2.17) should not be very surprising: if (2.4) holds with $s \neq \phi(s)$, then $f_t(\phi(s)) = h(s)^{-1} f_t(s)$, that is, the same relation (2.4) holds with s on its left-hand side replaced by $\phi(s)$, and $\phi(s)$ replaced by s. In other words, one may as well map $s' = \phi(s)$ by ϕ to $\phi(s') = s$ for the relation (2.4) to hold. Mapping $\phi(s)$ to s is equivalent to (2.17).

2.2 Rigidity of Integral Representations

We are interested in finding out when two $S\alpha S$ processes have the same law (finite-dimensional distributions), that is,

$$\Big\{\int_S f_t(s) M(ds)\Big\}_{t\in T} \stackrel{d}{=} \Big\{\int_{\widetilde{S}} \widetilde{f}_t(\widetilde{s}) \widetilde{M}(d\widetilde{s})\Big\}_{t\in T}, \tag{2.18}$$

where $\{f_t\} \in L^\alpha(S, \mathscr{B}(S), m)$, $\{\widetilde{f}_t\} \in L^\alpha(\widetilde{S}, \mathscr{B}(\widetilde{S}), \widetilde{m})$ and M ($\widetilde{M}$, resp.) is a $S\alpha S$ random measure on S ($\widetilde{S}$, resp.) with control measure m ($\widetilde{m}$, resp.). From an equivalent perspective, by studying (2.18), we are interested in the implications of expressing the same process through two different integral representations. Note that, in terms of the exponents of the characteristic functions of the two sides (cf. (1.24)), the relation (2.18) is equivalent to

[9] This weaker characterization will be used in the proof of Proposition 2.2. A null-isomorphism (mod 0) is defined before Theorem 2.1 below.

$$\Big\|\sum_{k=1}^n \theta_k f_{t_k}\Big\|_\alpha^\alpha = \int_S \Big|\sum_{k=1}^n \theta_k f_{t_k}(s)\Big|^\alpha m(ds) = \int_{\widetilde{S}} \Big|\sum_{k=1}^n \theta_k \widetilde{f}_{t_k}(\widetilde{s})\Big|^\alpha \widetilde{m}(d\widetilde{s}) = \Big\|\sum_{k=1}^n \theta_k \widetilde{f}_{t_k}\Big\|_\alpha^\alpha. \tag{2.19}$$

The relation (2.19) induces a natural *linear isometry*,[10]

$$U_0 : F = \{f_t, t \in T\} \to L^\alpha(\widetilde{S},\widetilde{m})$$

by

$$U_0 f_t = \widetilde{f}_t. \tag{2.20}$$

It turns out that linear isometries between L^α spaces have a very special structure. This is described in the celebrated Banach-Lamperti theorem in Section 2.2.1. This result concerns isometries defined on the entire space L^α, whereas the isometry (2.20) is defined in general only on a subspace. When working with minimal representations, however, the isometry (2.20) can be extended uniquely to one on the entire space L^α (Section 2.2.2 below). Thus, in the minimal case, the isometry (2.20) is characterized by the same special structure of isometries between L^α spaces appearing in the Banach-Lamperti theorem. The implications of the special structure on integral representations of stable processes are summarized and discussed in Section 2.2.3.

2.2.1 Banach-Lamperti Theorem and Its Extensions

The following is the celebrated result of Banach [4] and Lamperti [29] on the isometries of L^α spaces, where $\alpha \in (0,2)$. A *null-isomorphism* used below refers to a measurable, nonsingular, one-to-one, and onto map with a measurable inverse. While it may not be measure preserving, by being nonsingular, it preserves sets of measure zero ("null" refers to nonsingular). A null-isomorphism (*mod 0*) is a null-isomorphism with its inverse defined up to sets of measure zero.

Theorem 2.1. *(Banach-Lamperti) Let $\alpha \in (0,2)$, and*

$$U : L^\alpha(S,m) \to L^\alpha(\widetilde{S},\widetilde{m}) \tag{2.21}$$

be a linear isometry. Then, there are measurable maps $\phi : \widetilde{S} \to S$ and $h : \widetilde{S} \to \mathbb{R}$ such that, for any $f \in L^\alpha(S,m)$,

$$(Uf)(\widetilde{s}) = h(\widetilde{s}) f(\phi(\widetilde{s})) \quad \widetilde{m}\text{-a.e.} \tag{2.22}$$

[10] Given two normed vector spaces V and W, a linear isometry is a map $U : V \to W$ which is linear (that is, $U(af+g) = aU(f)+U(g)$ for any $f,g \in V$ and $a \in \mathbb{R}$) and preserves the norms (that is, $\|U(f)\| = \|f\|$). Note also that the isometry U_0 in (2.20) actually acts on the linear span span(F) and, more generally, on the closure of the linear span in $L^\alpha(S,m)$. We write just F for notational simplicity and the understanding that the values of U_0 on the span are determined by those on F from the definition of linear isometry.

and

$$dm = (|h|^{\alpha} d\widetilde{m}) \circ \phi^{-1}. \tag{2.23}$$

Conversely, (2.22) defines a linear isometry when ϕ and h satisfy (2.23).

Furthermore, if U is onto, then ϕ can be chosen as a null-isomorphism (mod 0) such that

$$\frac{d(m \circ \phi)}{d\widetilde{m}}(\widetilde{s}) = |h(\widetilde{s})|^{\alpha} > 0 \quad \widetilde{m}\text{-a.e.} \tag{2.24}$$

Conversely, if ϕ is a null-isomorphism (mod 0) satisfying (2.24), then (2.22) defines a linear isometry between $L^{\alpha}(S,m)$ and $L^{\alpha}(\widetilde{S},\widetilde{m})$.

Remark 2.2. The relation (2.24) is equivalent to

$$h(\widetilde{s}) = \varepsilon(\widetilde{s}) \Big\{ \frac{d(m \circ \phi)}{d\widetilde{m}}(\widetilde{s}) \Big\}^{1/\alpha} \quad \widetilde{m}\text{-a.e.} \tag{2.25}$$

for some $\varepsilon : \widetilde{S} \to \{-1,1\}$. We will often use this form in the presentation below.

Remark 2.3. The Banach-Lamperti Theorem 2.1 essentially states that linear isometries between L^{α} spaces can only be obtained by a change of variables. Indeed, in the L^{α} context, one should focus on integrals of the type $\int_S |f(s)|^{\alpha} dm(s)$ and $\int_{\widetilde{S}} |Uf(\widetilde{s})|^{\alpha} d\widetilde{m}(\widetilde{s})$. To go back from the "new" space $(\widetilde{S},\widetilde{m})$ to the "old" space (S,m), one must replace the variable $\widetilde{s}$ by the old one $s = \phi(\widetilde{s})$ and also scale the function f by the factor $h(\widetilde{s})$ to take care of the Jacobian of the transformation given in (2.24), namely $dm(s) = |h(\widetilde{s})|^{\alpha} d\widetilde{m}(\widetilde{s})$ with $s = \phi(\widetilde{s})$. One then has

$$\int_{\widetilde{S}} |Uf(\widetilde{s})|^{\alpha} d\widetilde{m}(\widetilde{s}) = \int_{\widetilde{S}} |h(\widetilde{s})|^{\alpha} |f(\phi(\widetilde{s}))|^{\alpha} |h(\widetilde{s})|^{-\alpha} dm(\phi(\widetilde{s})) = \int_S |f(s)|^{\alpha} dm(s).$$

Remark 2.4. The Banach-Lamperti theorem does not hold when $\alpha = 2$. Here is a simple counterexample. The $L^2(\mathbb{R})$-Fourier transform

$$\begin{aligned}
\widehat{f}(x) &= \frac{1}{\sqrt{2\pi}} \int_{\mathbb{R}} e^{ixu} f(u) du \\
&= \frac{1}{\sqrt{2\pi}} \int_{\mathbb{R}} \cos(xu) f(u) du + i \frac{1}{\sqrt{2\pi}} \int_{\mathbb{R}} \sin(xu) f(u) du \\
&=: \widehat{f}(1,x) + i\widehat{f}(2,x), \quad x \in \mathbb{R},
\end{aligned}$$

is defined for functions $f \in L^2(\mathbb{R}, du)$ so that $\int_{\mathbb{R}} |\widehat{f}(x)|^2 dx < \infty$ and Parseval's identity holds, namely,

$$\int_{\mathbb{R}} |f(u)|^2 du = \int_{\mathbb{R}} |\widehat{f}(x)|^2 dx = \int_{\mathbb{R}} (|\widehat{f}(1,x)|^2 + |\widehat{f}(2,x)|^2) dx \tag{2.26}$$

(e.g., Dym and McKean [12]). The Fourier transform leads naturally to a linear isometry of L^2 spaces. Indeed, let one measure space be $(S,m) = (\mathbb{R}, du)$ and the other measure space be $(\widetilde{S},\widetilde{m}) = (\{1,2\} \times \mathbb{R}, n(dj)dx)$, where $n(dj)$ denotes the counting measure on $\{1,2\}$. Define a map $U : L^2(\mathbb{R}, du) \to L^2(\{1,2\} \times \mathbb{R}, n(dj)dx)$

as

$$(Uf)(j,x) = \widehat{f}(j,x), \quad j = 1,2,\ x \in \mathbb{R}.$$

In view of (2.26), the map is a linear isometry from $L^2(\mathbb{R},du)$ to $L^2(\{1,2\} \times \mathbb{R}, n(dj)dx)$. But it clearly cannot satisfy the relation (2.22). For example, with $f_t = 1_{[0,t]} \in L^2(\mathbb{R},du)$, $t \in \mathbb{R}$, we have

$$\widehat{f_t}(x) = \frac{1}{\sqrt{2\pi}} \int_0^t e^{ixu} du = \frac{1}{\sqrt{2\pi}} \frac{e^{ixt}-1}{ix} = \frac{1}{\sqrt{2\pi}} \Big(\frac{\sin(tx)}{x} + i \frac{(1-\cos(tx))}{x} \Big)$$

and hence, e.g.,

$$\widehat{f_t}(1,x) = \frac{1}{\sqrt{2\pi}} \frac{\sin(tx)}{x}.$$

If the relation (2.22) was satisfied, it would imply: for fixed $t \in \mathbb{R}$,

$$\frac{1}{\sqrt{2\pi}} \frac{\sin(tx)}{x} = h(1,x) 1_{[0,t]}(\phi(1,x)) \quad \text{a.e. } dx \tag{2.27}$$

for some $h(1,\cdot) : \mathbb{R} \to \mathbb{R}$ and $\phi(1,\cdot) : \mathbb{R} \to \mathbb{R}$. By Lemma 1.1, (i), the relation (2.27) holds a.e. $dtdx$ and by Lemma 1.1, (ii), there is $x = x_0 \in \mathbb{R}$ ($x_0 \neq 0$), that the same relation (2.27) holds a.e. dt, that is,

$$\frac{1}{\sqrt{2\pi}} \frac{\sin(tx_0)}{x_0} = h(1,x_0) 1_{[0,t]}(\phi(1,x_0)) \quad \text{a.e. } dt. \tag{2.28}$$

The relation (2.28) does not hold: its left-hand side is (up to a constant) a sine function in t, and its right-hand side takes value $h(1,x_0)$ or 0, depending on whether $\phi(1,x_0) \in [0,t]$. Thus, we get a contradiction to the relation (2.22) when $\alpha = 2$. A stochastic process perspective on this example is provided in Remark 2.9.

Remark 2.5. The function h in Theorem 2.1 is unique $\widetilde{m}$-a.e. The function ϕ when restricted to the set

$$\widetilde{S}_0 = \text{supp}\{Uf : f \in L^\alpha(S,m)\}$$

is also unique $\widetilde{m}$-a.e. (Checking these facts is left as an exercise.)

PROOF: Suppose for simplicity that $\widetilde{m}(\widetilde{S}) < \infty$ and focus only on the first part of the theorem. The general case is treated in Section 7, Chapter 15 of Royden [50], and the second statement of the theorem is Problem 28 in Chapter 15 of Royden [50]. We work first with indicator functions.

The linear isometry (2.21) acts on functions, in particular, indicator functions. We can use it to define a mapping T which maps a set A in S into a set TA in $\widetilde{S}$ as follows. Define $T : S \to \widetilde{S}$ such that, for $A \in \mathscr{B}(S)$,

$$TA = \{\widetilde{s} : (U1_A)(\widetilde{s}) \neq 0\} = \text{supp}\{U1_A\} \in \mathscr{B}(\widetilde{S}).$$[11]

[11] As we will see below in (2.30), the mapping $T : S \to \widetilde{S}$ will play the role of $\phi^{-1} : \widetilde{S} \to S$. The map U, on the other hand, is defined as in (2.22) as the image in $\widetilde{S}$ of a function f in S. We first consider

If $A \cap B = \emptyset$, then

$$\|1_A + 1_B\|_\alpha^\alpha + \|1_A - 1_B\|_\alpha^\alpha = 2\|1_A\|_\alpha^\alpha + 2\|1_B\|_\alpha^\alpha.$$

Since U is an isometry, it follows that

$$\|U1_A + U1_B\|_\alpha^\alpha + \|U1_A - U1_B\|_\alpha^\alpha = 2\|U1_A\|_\alpha^\alpha + 2\|U1_B\|_\alpha^\alpha.$$

Therefore, by using Lemma 2.2, if $A \cap B = \emptyset$, then

$$U1_A U1_B = 0 \quad \widetilde{m}\text{-a.e.} \quad \text{and thus} \quad T(A+B) = TA + TB \quad \widetilde{m}\text{-a.e.}, \tag{2.29}$$

where $+$ is used to indicate the union of disjoint sets. More generally, one can show that

1. $T(\sum_{i=1}^\infty A_i) = \sum_{i=1}^\infty T(A_i)$ $\widetilde{m}$-a.e. for mutually disjoint A_i,
2. $T(S \setminus A) = TS \setminus TA$ $\widetilde{m}$-a.e.,
3. $\widetilde{m}(TA) = 0 \Leftrightarrow m(A) = 0$.

(Showing that the mapping T has indeed the properties 1–3 is left as an exercise.) Such T is called a regular set isomorphism. It is known (Theorem 32.5 in Sikorski [58], Proposition 2.1 in Rosiński [44]) that T is induced by a point transformation, that is, there is $\phi : \widetilde{S} \to S$ such that

$$TA = \phi^{-1}(A) \quad \widetilde{m}\text{-a.e.} \tag{2.30}$$

Let now

$$h(\widetilde{s}) = (U1_S)(\widetilde{s}) = (U1_A)(\widetilde{s}) + (U1_{S\setminus A})(\widetilde{s}).$$

By (2.29), $(U1_A)(\widetilde{s})$ and $(U1_{S\setminus A})(\widetilde{s})$ have disjoint supports $\widetilde{m}$-a.e. Since the support of $(U1_A)(\widetilde{s})$ is denoted TA, it follows that by expressing $U1_A(\widetilde{s})$ as $h(\widetilde{s})1_{TA}(\widetilde{s})$,

$$U1_A(\widetilde{s}) = h(\widetilde{s})1_{TA}(\widetilde{s}) = h(\widetilde{s})1_{\phi^{-1}(A)}(\widetilde{s}) = h(\widetilde{s})1_A(\phi(\widetilde{s})) \quad \widetilde{m}\text{-a.e.}$$

which is (2.22) in the special case $f = 1_A$. This last relation can be extended by approximation to

$$Uf(\widetilde{s}) = h(\widetilde{s})f(\phi(\widetilde{s})) \quad \widetilde{m}\text{-a.e.},$$

which proves (2.22).

Finally,

$$m(A) = \|1_A\|_\alpha^\alpha = \|U1_A\|_\alpha^\alpha = \int_{\widetilde{S}} |h(\widetilde{s})|^\alpha 1_{TA}(\widetilde{s})^\alpha \widetilde{m}(d\widetilde{s}) = \int_{\phi^{-1}(A)} |h(\widetilde{s})|^\alpha \widetilde{m}(d\widetilde{s}),$$

by (2.30). This proves (2.23). □

The following two auxiliary results were used in the proof above.

---

indicator functions $f = 1_A$, $A \in \mathscr{B}(S)$. The map T maps A into the image in $\widetilde{S}$ of 1_A, excluding the points where this image vanishes.

Lemma 2.1. *Let $\alpha \in (0,2)$ and $u, v \in \mathbb{R}$. Then,*

$$|u+v|^\alpha + |u-v|^\alpha \leq 2|u|^\alpha + 2|v|^\alpha \tag{2.31}$$

and the equality holds if and only if $uv = 0$.

PROOF: We may suppose without loss of generality that $u, v \geq 0$ or even more specifically that $u > v \geq 0$. By dividing (2.31) by u and setting $z = v/u \in [0,1)$, the inequality (2.31) is equivalent to

$$(1+z)^\alpha + (1-z)^\alpha \leq 2 + 2z^\alpha.$$

Set

$$g(z) = (1+z)^\alpha + (1-z)^\alpha - 2z^\alpha - 2.$$

Note that $g(0) = 0$ and $g(1) = 2^\alpha - 2^2 < 0$. It is then enough to show that $g'(z) < 0$ for $z \in (0,1)$. Indeed,

$$g'(z) = \alpha((1+z)^{\alpha-1} - (1-z)^{\alpha-1} - 2z^{\alpha-1}) = \alpha z^{\alpha-1} h(z^{-1}),$$

where

$$h(u) = (u+1)^{\alpha-1} - (u-1)^{\alpha-1} - 2, \quad u > 1.$$

It remains to observe that $h(u) < 0$ for $u > 1$. When $\alpha \in (1,2)$ or $\alpha - 1 \in (0,1)$, this follows from

$$(u+1)^{\alpha-1} = (2+u-1)^{\alpha-1} \leq 2^{\alpha-1} + (u-1)^{\alpha-1} < 2 + (u-1)^{\alpha-1}.$$

When $\alpha = 1$, $h(u) = -2 < 0$. When $\alpha \in (0,1)$, this follows from $(u+1)^{\alpha-1} < (u-1)^{\alpha-1}$ since $u - 1 < u + 1$ and $\alpha - 1 < 0$. □

Lemma 2.2. *Let $\alpha \in (0,2)$ and $f, g \in L^\alpha(S,m)$. Then,*

$$\|f+g\|_\alpha^\alpha + \|f-g\|_\alpha^\alpha \leq 2\|f\|_\alpha^\alpha + 2\|g\|_\alpha^\alpha$$

and the equality holds if and only if $fg = 0$ m-a.e.

PROOF: The inequality in the lemma follows by taking $u = f(s)$ and $v = g(s)$ in Lemma 2.1 and integrating both sides of (2.31) over S. The equality holds if and only if $fg = 0$ m-a.e. by using the second part of Lemma 2.1. □

Theorem 2.1 concerns linear isometries defined on entire spaces L^α, $\alpha \in (0,2)$. The isometry (2.20) underlying (2.18), on the other hand, is defined on a specific subspace of L^α, namely, the linear span generated by $f_t, t \in T$. The following result, Theorem 2.2, due to Rosiński [44], extends Theorem 2.1 to any linear isometry defined on a subspace of L^α. Hence we shall now suppose only that the functions in F form a subspace of $L^\alpha(S, \mathscr{B}(S), m)$.

Suppose then that F is a subspace of $L^\alpha(S,\mathscr{B}(S),m)$. By Lemma 3.4 in Hardin [19], there is a function $f^* \in \overline{\mathrm{span}}(F)_\alpha$, the closure in $L^\alpha(S,m)$ of the linear span of F, such that

$$\mathrm{supp}\{f^*\} = \mathrm{supp}\{F\} \quad m\text{-a.e.} \tag{2.32}$$

In other words, there is a function, we call it f^*, which belongs to the closure of the span of F whose support coincides with the support of the collection F (the support of a collection of functions is defined following (1.25)).

Example 2.4. Let $F = \{f_t\}_{t\in T} \subset L^\alpha(S,m)$. Then, in view of (2.32), there is a function f^* belonging to the closure of the linear span generated by f_t, $t \in T$, whose support is identical to the support of F (m-a.e.).

Theorem 2.2. *(Rosiński) Let $\alpha \in (0,2)$ and $F \subset L^\alpha(S,\mathscr{B}(S),m)$ be such that*

$$\sigma\{f/f^*,\ f\in F\} \subset \mathscr{A} \subset \overline{\sigma}\{f/f^*,\ f\in F\} \tag{2.33}$$

for a σ-field $\mathscr{A}$ and where $\overline{\sigma}\{f/f^*,\ f\in F\}$ is the smallest σ-field containing $\sigma\{f/f^*,\ f\in F\}$ and m-null sets from $\mathscr{B}(S)$. Let also

$$U_0 : F \to L^\alpha(\widetilde{S},\widetilde{m})$$

be a linear isometry. Then, there are measurable maps $\phi : \widetilde{S} \to S$ and $h : \widetilde{S} \to \mathbb{R}$ such that, for any $f \in F$,

$$(U_0 f)(\widetilde{s}) = h(\widetilde{s}) f(\phi(\widetilde{s})) \quad \widetilde{m}\text{-a.e.} \tag{2.34}$$

and

$$|f^*|^\alpha dm = |f^*|^\alpha \left((|h|^\alpha d\widetilde{m}) \circ \phi^{-1} \right) \quad \text{on } \mathscr{A}. \tag{2.35}$$

Conversely, if (2.35) holds, then (2.34) defines an isometry on F.

Remark 2.6. Here, the relation (2.34) holds for f belonging to the subspace F of L^α. In (2.35), one is restricted to the support of F because $f^* = 0$ outside the support.

Remark 2.7. The theorem says in particular that any linear isometry in L^α entails a relation of the form (2.34). When a linear isometry is induced by a *minimal* representation, we will see below (cf. the proof of Proposition 2.2) that $\mathscr{A}$ can be taken to be $\mathscr{B}(S)$. In this case, we will be able to cancel out the function f^* on both sides of the relation (2.35).

PROOF OUTLINE: In establishing (2.34), we shall use Rudin's theorem (Rudin [51]) which states the following: If $p > 0$ is not an even integer, if ν and $\widetilde{\nu}$ are finite measures on X and $\widetilde{X}$, respectively, and if $g_1,\dots,g_n \in L^p(X,\nu)$ and $\widetilde{g}_1,\dots,\widetilde{g}_n \in L^p(\widetilde{X},\widetilde{\nu})$ are such that

$$\int_X \Big|1 + \sum_{i=1}^n \lambda_i g_i\Big|^p d\nu = \int_{\widetilde{X}} \Big|1 + \sum_{i=1}^n \lambda_i \widetilde{g}_i\Big|^p d\widetilde{\nu} \tag{2.36}$$

for all $\lambda_1,\dots,\lambda_n \in \mathbb{R}$, then

$$\nu \circ (g_1,\dots,g_n)^{-1} = \widetilde{\nu} \circ (\widetilde{g}_1,\dots,\widetilde{g}_n)^{-1}. \tag{2.37}$$

The last property is called the *equimeasurability* of $(g_1,\dots,g_n)$ and $(\widetilde{g}_1,\dots,\widetilde{g}_n)$. Rudin's theorem can be viewed as a generalization of the Banach-Lamperti theorem, Theorem 2.1 above (see, for example, the discussion in the introduction of Hardin [19]). In particular, our assumption of linear isometry of U_0 will have to be reduced to the form (2.36). To achieve this goal, two issues will need to be addressed. First, the linear isometry property of U_0 implies that

$$\int_S \Big|\sum_{i=0}^n \lambda_i f_i\Big|^p dm = \int_{\widetilde{S}} \Big|\sum_{i=0}^n \lambda_i U_0 f_i\Big|^p d\widetilde{m}, \tag{2.38}$$

for any $\lambda_j \in \mathbb{R}$ and $f_i \in F$. To obtain (2.36) from (2.38), we shall take $\lambda_0 = 1$ and $f_0 = f^*$, where f^* defined in (2.32) has full support. Then, f^* can be factored out of the sum $\sum_{i=0}^n \lambda_i f_i$ as $f^*(1+\sum_{i=0}^n \lambda_i f_i/f^*)$ and the relation (2.38) can be reduced to the form (2.36). Second, the collection of functions F can be uncountable, while the condition (2.36) involves a finite number of functions. We shall deal with the uncountability issue by choosing a suitable countable basis of F. We next proceed to outlining the steps of this approach.

By the separability of $L^\alpha(S,\mathscr{B}(S),m)$, one can choose a countable collection $\{f_i\}_{i\geq 1} \subset F$ which is dense in F. Then,

$$\overline{\sigma}\{f_i/f^*,\ i \geq 1\} = \overline{\sigma}\{f/f^*,\ f \in F\}.$$

Put $\widetilde{f}_i = U_0 f_i$, $\widetilde{f}^* = U_0 f^*$, $S_0 = \mathrm{supp}\{f^*\}$ and $\widetilde{S}_0 = \mathrm{supp}\{\widetilde{f}^*\}$. By using Lemma 3.4 in Hardin [19], one has $\widetilde{S}_0 = \mathrm{supp}\{\widetilde{f}_i,\ i \geq 1\}$. Set

$$F = \Big(\frac{f_1}{f^*}, \frac{f_2}{f^*}, \dots\Big) \quad \text{on } S_0, \quad \widetilde{F} = \Big(\frac{\widetilde{f}_1}{\widetilde{f}^*}1_{\widetilde{S}_0}, \frac{\widetilde{f}_2}{\widetilde{f}^*}1_{\widetilde{S}_0}, \dots\Big).$$

Set also $d\nu = |f^*|^\alpha dm$ and $d\widetilde{\nu} = |\widetilde{f}^*|^\alpha d\widetilde{m}$. Since U_0 is an isometry, we have, for $n \geq 1$ and $\lambda_1,\dots,\lambda_n \in \mathbb{R}$,

$$\int_{S_0} \Big|1+\sum_{i=1}^n \lambda_i \frac{f_i}{f^*}(s)\Big|^\alpha \nu(ds) = \int_S \Big|f^*(s)+\sum_{i=1}^n \lambda_i f_i(s)\Big|^\alpha m(ds)$$

$$= \int_{\widetilde{S}} \Big|\widetilde{f}^*(\widetilde{s})+\sum_{i=1}^n \lambda_i \widetilde{f}_i(\widetilde{s})\Big|^\alpha \widetilde{m}(d\widetilde{s}) = \int_{\widetilde{S}_0} \Big|1+\sum_{i=1}^n \lambda_i \frac{\widetilde{f}_i}{\widetilde{f}^*}(\widetilde{s})\Big|^\alpha \widetilde{\nu}(d\widetilde{s}). \tag{2.39}$$

By Rudin's theorem given above (see (2.36) and (2.37)), since (2.39) holds for arbitrary n, we conclude that

$$\nu \circ F^{-1} = \widetilde{\nu} \circ \widetilde{F}^{-1},$$

which is the important step. This relation implies (Theorem 2.1 in Rosiński [44]) that there is $\phi : \widetilde{S} \to S_0$ such that $\nu = \widetilde{\nu} \circ \phi^{-1}$ on $\mathscr{A}$ and

$$\widetilde{F}(\widetilde{s}) = F(\phi(\widetilde{s}))$$

$\widetilde{\nu}$-a.e. or, equivalently, for any $i \geq 1$,

$$\frac{\widetilde{f_i}}{\widetilde{f^*}}(\widetilde{s}) = \frac{f_i}{f^*}(\phi(\widetilde{s}))$$

$\widetilde{\nu}$-a.e. By setting

$$h(\widetilde{s}) = \begin{cases} \widetilde{f^*}(\widetilde{s})/f^*(\phi(\widetilde{s})), & \text{if } \widetilde{s} \in \widetilde{S}_0, \\ 0, & \text{otherwise,} \end{cases}$$

the previous relation becomes

$$\widetilde{f_i}(\widetilde{s}) = h(\widetilde{s}) f_i(\phi(\widetilde{s})) \quad \widetilde{\nu}\text{-a.e.}$$

Since $\widetilde{m}$ is equivalent to $\widetilde{\nu}$ on $\widetilde{S}_0$ and $\widetilde{f_i} = 0$ $\widetilde{\nu}$-a.e. on $\widetilde{S} \setminus \widetilde{S}_0$, we have, for $i \geq 1$,

$$\widetilde{f_i}(\widetilde{s}) = h(\widetilde{s}) f_i(\phi(\widetilde{s})) \quad \widetilde{m}\text{-a.e.}$$

Since $\widetilde{f_i} = U_0 f_i$, we obtain

$$(U_0 f_i)(\widetilde{s}) = h(\widetilde{s}) f_i(\phi(\widetilde{s})) \quad \widetilde{m}\text{-a.e.}$$

The relation (2.34) follows by an approximation argument since $\{f_i\}$ is dense in F. Finally, the relation (2.35) can be proved by using Theorem 4.3 in Rosiński [44]. □

2.2.2 *Minimal Representations and Isometries*

The next result provides another important characterization of minimal integral representations. Its proof follows Rosiński [48].

Proposition 2.2. *An integral representation $F = \{f_t\}_{t \in T} \subset L^\alpha(S, \mathscr{B}(S), m)$ is minimal if and only if every linear isometry*

$$U_0 : F = \{f_t\}_{t \in T} \to L^\alpha(\widetilde{S}, \mathscr{B}(\widetilde{S}), \widetilde{m})$$

has a unique extension to a linear isometry

$$U : L^\alpha(S, \mathscr{B}(S), m) \to L^\alpha(\widetilde{S}, \mathscr{B}(\widetilde{S}), \widetilde{m}).$$

PROOF: Suppose first that the integral representation $\{f_t\}_{t\in T}$ is minimal. By Definition 2.1, it satisfies the condition (2.3). If f^* is a function as in (2.32), then (2.3) implies that

$$\sigma\{f_t/f^*,\, t\in T\} = \mathscr{B}(S) \quad m\text{-a.e.} \tag{2.40}$$

Indeed, since

$$\frac{f_u}{f_v} = \left(\frac{f_u}{f^*}\right)\left(\frac{f_v}{f^*}\right)^{-1},$$

we have

$$\sigma\{f_u/f_v\} \subset \sigma\{f_t/f^*,\, t\in T\}$$

and hence

$$\mathscr{B}(S) = \sigma\{f_u/f_v,\, u,v\in T\} \subset \sigma\{f_t/f^*,\, t\in T\}.$$

Now, the relation (2.40) shows that Theorem 2.2 applies with $\mathscr{A} = \mathscr{B}(S)$. With this choice of $\mathscr{A}$, since $|f^*|^\alpha$ is $\mathscr{B}(S)$-measurable and has a support on $S = \text{supp}\{F\}$, $|f^*|^\alpha$ can be canceled on both sides of (2.35). Then, (2.34)–(2.35) become (2.22)–(2.23) and define a linear isometry on the entire space $L^\alpha(S,m)$. The linear isometry is unique by Remark 2.5.

We now turn to the converse. We assume that every map U_0 has a unique extension U as stated in the proposition. We want to show minimality. To do so, we use the criterion stated in Remark 2.1. Suppose (2.4) holds with $\phi\circ\phi = Id$, where $\phi: S\to S$ is a null-isomorphism (mod 0). We need to show that $\phi = Id$ m-a.e. Let f^* be a function as in (2.32) where $\text{supp}\{F\} = S$ m-a.e. Note that, by dividing (2.4) by (2.4) with f_t replaced by f^*, the function h cancels out and we obtain

$$f_t/f^* = (f_t/f^*)\circ\phi \quad m\text{-a.e.} \tag{2.41}$$

Let $dm^* = |f^*|^\alpha dm$ and

$$\nu = \frac{1}{2}m^* + \frac{1}{2}m^*\circ\phi^{-1}. \tag{2.42}$$

Consider the special map $U_0: F\to L^\alpha(S,\nu)$ given by

$$U_0 f_t := f_t/f^*. \tag{2.43}$$

Note that U_0 is a linear isometry since, for $f = \sum_{k=1}^n \theta_k f_{t_k}$,

$$\|U_0 f\|_\alpha^\alpha = \frac{1}{2}\int_S \left|\frac{f}{f^*}\right|^\alpha dm^* + \frac{1}{2}\int_S \left|\frac{f}{f^*}\circ\phi^{-1}\right|^\alpha dm^*$$

$$= \int_S \left|\frac{f}{f^*}\right|^\alpha dm^* = \int_S |f|^\alpha dm = \|f\|_\alpha^\alpha$$

by using (2.41). By the assumption of the converse of our proposition, there is an isometry

$$U: L^\alpha(S,m)\to L^\alpha(S,\nu)$$

extending U_0. The idea is to introduce later another map U_1 defined on $L^\alpha(S,m)$ which is also an extension of U_0, and then to identify U and U_1 because the extension of U_0 is unique.

Since U is defined on the whole of $L^\alpha(S,m)$, we can now apply the Banach-Lamperti Theorem 2.1 and conclude that there is $\psi : S \to S$ such that

$$Uf = h \cdot f \circ \psi$$

and

$$dm = (|h|^\alpha d\nu) \circ \psi^{-1}. \tag{2.44}$$

Since U extends U_0, we can apply the relation (2.43) to get $Uf^* = f^*/f^* = 1$. Then, $1 = Uf = h \cdot f^* \circ \psi$ and hence

$$h = \frac{1}{f^* \circ \psi}. \tag{2.45}$$

Since $\phi \circ \phi = Id$, the definition (2.42) of ν implies that

$$\nu \circ \phi^{-1} = \frac{1}{2} m^* \circ \phi^{-1} + \frac{1}{2} m^* \circ \phi^{-1} \circ \phi^{-1} = \frac{1}{2} m^* \circ \phi^{-1} + \frac{1}{2} m^* = \nu,$$

that is, the measure ν is ϕ-invariant. Then, the map $U_1 : L^\alpha(S,m) \to L^\alpha(S,\nu)$ defined by

$$U_1 f = (h \circ \phi) \cdot f \circ (\psi \circ \phi) \tag{2.46}$$

is also an isometry. Indeed, by using ϕ-invariance of ν and the fact that U is an isometry, observe that

$$\int_S |U_1 f|^\alpha d\nu = \int_S |h \circ \phi|^\alpha |f \circ (\psi \circ \phi)|^\alpha d\nu$$

$$= \int_S |h|^\alpha |f \circ \psi|^\alpha d\nu = \int_S |Uf|^\alpha d\nu = \int_S |f|^\alpha dm.$$

The map U_1 also extends U_0. Indeed, since U extends U_0 and by using (2.41) and (2.43), we have

$$U_1 f_t = (h \cdot f_t \circ \psi) \circ \phi = (Uf_t) \circ \phi = (U_0 f_t) \circ \phi = \frac{f_t}{f^*} \circ \phi = \frac{f_t}{f^*} = U_0 f_t.$$

But the extension U of U_0 is assumed to be unique. Therefore $U = U_1$ and hence

$$h \cdot f \circ \psi = (h \circ \phi) \cdot f \circ (\psi \circ \phi).$$

Thus, by Remark 2.5, we get $h = h \circ \phi$ m-a.e. and also

$$\psi \circ \phi = \psi \quad m\text{-a.e.} \tag{2.47}$$

Observe that, by using (2.44), (2.45), (2.42), and (2.47),

$$m(A) = \int_S (1_A \circ \psi)|h|^\alpha d\nu = \int_S (1_A \circ \psi)|f^* \circ \psi|^{-\alpha} d\nu$$

$$= \frac{1}{2}\int_S (1_A \circ \psi)|f^* \circ \psi|^{-\alpha} dm^* + \frac{1}{2}\int_S (1_A \circ \psi \circ \phi)|f^* \circ \psi \circ \phi|^{-\alpha} dm^*$$

$$= \frac{1}{2}\int_S (1_A \circ \psi)|f^* \circ \psi|^{-\alpha} dm^* + \frac{1}{2}\int_S (1_A \circ \psi \circ \phi)|f^* \circ \psi|^{-\alpha} dm^*$$

$$= \int_S (1_A \circ \psi)|f^* \circ \psi|^{-\alpha} dm^*$$

and hence that

$$dm = |f^*|^{-\alpha} d(m^* \circ \psi^{-1}).$$

This ensures that

$$V_1 : L^\alpha(S,m) \to L^\alpha(S,m^*),$$

defined by

$$V_1 f = (f/f^*) \circ \psi,$$

is an isometry. Indeed,

$$\int_S |V_1 f|^\alpha dm^* = \int_S |f \circ \psi|^\alpha |f^* \circ \psi|^{-\alpha} dm^* = \int_S |f|^\alpha |f^*|^{-\alpha} d(m^* \circ \psi^{-1}) = \int_S |f|^\alpha dm.$$

Similarly,

$$V_2 : L^\alpha(S,m) \to L^\alpha(S,m^*),$$

defined by

$$V_2 f = f/f^*,$$

is also an isometry. Note that V_1 extends U_0 since by (2.45) and the fact that U extends U_0,

$$V_1 f_t = \left(\frac{f_t}{f^*}\right) \circ \psi = h \cdot (f_t \circ \phi) = U f_t = U_0 f_t.$$

Similarly, by using (2.43), V_2 also extends U_0. By the uniqueness of the extension, $V_1 = V_2$ and hence $\psi = Id$ m-a.e. (see Remark 2.5). Then, by (2.47), $\phi(s) = s$ for $s \in A$ and $\phi(s) \in A$ with $m(A^c) = 0$, that is, $\phi = Id$ on $A \cap \phi^{-1}(A)$. Since ϕ is nonsingular and $m(A^c) = 0$, we have $m(\phi^{-1}(A^c)) = 0$ and hence $m((A \cap \phi^{-1}(A))^c) = m(A^c \cup \phi^{-1}(A^c)) \le m(A^c) + m(\phi^{-1}(A^c)) = 0$ and hence $\phi = Id$ m-a.e. □

2.2.3 Rigidity Properties

We gather here consequences of the results of Sections 2.2.1 and 2.2.2 for the relation (2.18). The first result concerns the case when one of the representations in (2.18) is minimal.

Proposition 2.3. *Suppose that the relation (2.18) holds and that the representation* $\{f_t\}$ *is minimal. Then, there are*

$$\phi : \widetilde{S} \to S \quad \text{and} \quad h : \widetilde{S} \to \mathbb{R}$$

such that, for all $t \in T$,

$$\widetilde{f}_t(\widetilde{s}) = h(\widetilde{s}) f_t(\phi(\widetilde{s})) \quad \widetilde{m}\text{-a.e.} \tag{2.48}$$

and

$$dm = (|h|^\alpha d\widetilde{m}) \circ \phi^{-1}. \tag{2.49}$$

PROOF: The map U_0 defined by (2.20) is an isometry on $\{f_t,\ t \in T\}$ as shown in (2.19). By Proposition 2.2, since $\{f_t\}$ is minimal, it can be extended uniquely to a linear isometry

$$U : L^\alpha(S,m) \to L^\alpha(\widetilde{S},\widetilde{m}).$$

The result now follows from Theorem 2.1. □

The second result concerns the case when both representations in (2.18) are minimal.

Proposition 2.4. *Suppose that the relation (2.18) holds and that the representations* $\{f_t\}$ *and* $\{\widetilde{f}_t\}$ *are minimal. Then, there are*

$$\text{null-isomorphism (mod 0) } \phi : \widetilde{S} \to S \quad \text{and} \quad \varepsilon : \widetilde{S} \to \{-1,1\}$$

such that, for all $t \in T$,

$$\widetilde{f}_t(\widetilde{s}) = \varepsilon(\widetilde{s}) \left\{ \frac{d(m \circ \phi)}{d\widetilde{m}}(\widetilde{s}) \right\}^{1/\alpha} f_t(\phi(\widetilde{s})) \quad \widetilde{m}\text{-a.e.} \tag{2.50}$$

PROOF: As in the proof of Proposition 2.3, consider the isometry U_0 and its unique extension U on $L^\alpha(S,m)$. Similarly, since $\{\widetilde{f}_t,\ t \in T\}$ is minimal, consider the isometry $\widetilde{U}_0(\widetilde{f}_t) = f_t$ and its unique extension $\widetilde{U} : L^\alpha(\widetilde{S},\widetilde{m}) \to L^\alpha(S,m)$. Then, $U \circ \widetilde{U}$ is an isometry from $L^\alpha(\widetilde{S},\widetilde{m})$ to $L^\alpha(\widetilde{S},\widetilde{m})$ satisfying $(U \circ \widetilde{U})(\widetilde{f}_t) = \widetilde{f}_t$. Since the identity map $I(f) = f$ is another such isometry from $L^\alpha(\widetilde{S},\widetilde{m})$ to $L^\alpha(\widetilde{S},\widetilde{m})$ and $\{\widetilde{f}_t,\ t \in T\}$ is minimal, the uniqueness of the extension (see Proposition 2.2) yields $U \circ \widetilde{U} = Id$. This shows that U is onto. Hence, by Theorem 2.1, ϕ in (2.48) can be chosen as a null-isomorphism (mod 0) and h in (2.48) satisfies (2.24) and hence can be written as

$$h(\widetilde{s}) = \varepsilon(\widetilde{s}) \left\{ \frac{d(m \circ \phi)}{d\widetilde{m}}(\widetilde{s}) \right\}^{1/\alpha}$$

$\widetilde{m}$-a.e. for some $\varepsilon : \widetilde{S} \to \{-1,1\}$. □

Finally, the last result concerns the case when neither representation in (2.18) is minimal. This case is not covered by the results presented in Sections 2.2.1 and 2.2.2 but can be proved using similar ideas (Rosiński [44], Theorem 3.1).

Proposition 2.5. *Suppose that the relation (2.18) holds. Then, there are $\phi : \widetilde{S} \to S$ and $h : \widetilde{S} \to \mathbb{R}$ such that*

$$\widetilde{f}_t(\widetilde{s}) = h(\widetilde{s}) f_t(\phi(\widetilde{s})) \quad a.e.\ \sigma(dt)\widetilde{m}(d\widetilde{s}), \tag{2.51}$$

where $\sigma(dt)$ is a σ-finite measure on the index space.

PROOF: Let F ($\widetilde{F}$, resp.) be the closure in $L^\alpha(S,m)$ ($L^\alpha(\widetilde{S},\widetilde{m})$, resp.) of the linear span of $\{f_t\}_{t\in T}$ ($\{\widetilde{f}_t\}_{t\in T}$, resp.). By (2.18)–(2.19), the mapping $U_0 f_t = \widetilde{f}_t$ extends naturally to a linear isometry of F onto $\widetilde{F}$. By Lemma 3.1 in Rosiński [44], there is $f^* \in F$ such that $\mathrm{supp}\{f^*\} = \mathrm{supp}\{F\}$ m-a.e. and $\mathrm{supp}\{\widetilde{f}^*\} = \mathrm{supp}\{\widetilde{F}\}$ $\widetilde{m}$-a.e., where $\widetilde{f}^* = U_0 f^*$. Let $S_0 = \{f^* \neq 0\}$ ($= \mathrm{supp}\{f^*\}$ a.e.) and $\widetilde{S}_0 = \{\widetilde{f}^* \neq 0\}$ ($= \mathrm{supp}\{\widetilde{f}^*\}$ a.e.). Assume without loss of generality that σ is a finite measure on the index space $(T, \mathscr{B}(T))$.

Define Borel maps $\zeta : S_0 \to L^0(T,\mathscr{B}(T),\sigma)$ and $\widetilde{\zeta} : \widetilde{S}_0 \to L^0(T,\mathscr{B}(T),\sigma)$ (with L^0 indicating the collection of measurable functions on the corresponding space) by

$$\zeta(s) = \left(\frac{f_t(s)}{f^*(s)}\right)_{t\in T}, \quad \widetilde{\zeta}(s) = \left(\frac{\widetilde{f}_t(\widetilde{s})}{\widetilde{f}^*(\widetilde{s})} 1_{\{\widetilde{s}\in\widetilde{S}_0\}}\right)_{t\in T}, \tag{2.52}$$

that is, for each fixed s and $\widetilde{s}$, $\zeta(s)$ and $\widetilde{\zeta}(\widetilde{s})$ are measurable functions on T. Let $\nu(ds) = |f^*(s)|^\alpha m(ds)$ and $\widetilde{\nu}(d\widetilde{s}) = |\widetilde{f}^*(\widetilde{s})|^\alpha \widetilde{m}(d\widetilde{s})$. Since U_0 defined above is a linear isometry, note that, for every $n \geq 1$ and $t_1,\dots,t_n \in T$, $\lambda_1,\dots,\lambda_n \in \mathbb{R}$,

$$\int_{S_0} \left|1 + \sum_{i=1}^n \lambda_i \frac{f_{t_i}}{f^*}(s)\right|^\alpha \nu(ds) = \int_S \left|f^*(s) + \sum_{i=1}^n \lambda_i f_{t_i}(s)\right|^\alpha m(ds) = \left\|f^* + \sum_{i=1}^n \lambda_i f_{t_i}\right\|^\alpha_{L^\alpha(S,m)}$$

$$= \left\|U_0(f^* + \sum_{i=1}^n \lambda_i f_{t_i})\right\|^\alpha_{L^\alpha(\widetilde{S},\widetilde{m})} = \left\|\widetilde{f}^* + \sum_{i=1}^n \lambda_i \widetilde{f}_{t_i}\right\|^\alpha_{L^\alpha(\widetilde{S},\widetilde{m})}$$

$$= \int_{\widetilde{S}} \left|\widetilde{f}^*(\widetilde{s}) + \sum_{i=1}^n \lambda_i \widetilde{f}_{t_i}(\widetilde{s})\right|^\alpha \widetilde{m}(d\widetilde{s}) = \int_{\widetilde{S}_0} \left|1 + \sum_{i=1}^n \lambda_i \frac{\widetilde{f}_{t_i}}{\widetilde{f}^*}(\widetilde{s})\right|^\alpha \widetilde{\nu}(d\widetilde{s}).$$

By Rudin's theorem stated in (2.36)–(2.37), we then have for every $n \geq 1$, $t_1,\dots,t_n \in T, B \in \mathscr{B}(\mathbb{R}^n)$,

$$\nu\left(S_0 \cap \left\{\left(\frac{f_{t_1}}{f^*},\dots,\frac{f_{t_n}}{f^*}\right) \in B\right\}\right) = \widetilde{\nu}\left(\widetilde{S}_0 \cap \left\{\left(\frac{\widetilde{f}_{t_1}}{\widetilde{f}^*},\dots,\frac{\widetilde{f}_{t_n}}{\widetilde{f}^*}\right) \in B\right\}\right). \tag{2.53}$$

Since $\widetilde{\nu}(\widetilde{S}\setminus\widetilde{S}_0) = 0$ and the sets $\{(g_{t_1},\dots,g_{t_n}) \in B\}$ generate the Borel σ-field associated with the space $L^0(T,\mathscr{B}(T),\sigma)$, we conclude from (2.53) that

$$\nu \circ \zeta^{-1} = \widetilde{\nu} \circ \widetilde{\zeta}^{-1}.$$

By Theorem 2.1 in Rosiński [44], there is $\phi : \widetilde{S} \to S_0$ such that

$$\widetilde{\zeta}(\widetilde{s}) = \zeta(\phi(\widetilde{s})) \quad \text{a.e. } \widetilde{\nu}(d\widetilde{s}). \tag{2.54}$$

Recall from (2.52) that the two hand sides of (2.54) are functions on T, which by (2.54) are equal in $L^0(T, \mathscr{B}(\mathbb{R}), \sigma)$ (that is, they coincide a.e. $\sigma(dt)$). By Fubini's theorem,

$$\Big(\widetilde{\zeta}(\widetilde{s})\Big)(t) = \Big(\zeta(\phi(\widetilde{s}))\Big)(t) \quad \text{a.e. } \sigma(dt)\nu(d\widetilde{s}) \tag{2.55}$$

or, by (2.52),

$$\widetilde{f}_t(\widetilde{s}) = \frac{\widetilde{f}^*(\widetilde{s})}{f^*(\phi(\widetilde{s}))} f_t(\phi(\widetilde{s})) \quad \text{a.e. } \sigma(dt)\nu(d\widetilde{s}). \tag{2.56}$$

Set $h(\widetilde{s}) = \widetilde{f}^*(\widetilde{s})/f^*(\phi(\widetilde{s}))$ if $\widetilde{s} \in \widetilde{S}_0$ and $= 0$ otherwise. Since $\widetilde{m}$ is equivalent to $\widetilde{\nu}$ on $\widetilde{S}_0$, the relation (2.56) implies (2.51) on $T \times \widetilde{S}_0$. Since $\widetilde{f}_t(\widetilde{s}) = 0$ a.e. $\sigma(dt)\widetilde{m}(d\widetilde{s})$ on $T \times (\widetilde{S} \setminus \widetilde{S}_0)$, the relation (2.51) then also holds on $T \times (\widetilde{S} \setminus \widetilde{S}_0)$. This shows that (2.51) holds on all $T \times \widetilde{S}$. □

Remark 2.8. In general, one cannot conclude that the relation in (2.51) holds a.e. $\widetilde{m}(d\widetilde{s})$ for each fixed $t \in T$. See Remark 3.1 in Rosiński [44] for a simple counterexample. But sufficient conditions on the representations can be provided for the relation in (2.51) to hold a.e. $\widetilde{m}(d\widetilde{s})$ for $t \in T$. For example, this is the case (Theorem 4.2 in Rosiński [44]) if the representations $\{f_t\}_{t\in T} \subset L^\alpha(S, \mathscr{B}(S), m)$ and $\{\widetilde{f}_t\}_{t\in T} \subset L^\alpha(\widetilde{S}, \mathscr{B}(\widetilde{S}), \widetilde{m})$ are *jointly separated*, that is, if there are a countable set $T_0 \subset T$ and null sets $N \subset S$, $\widetilde{N} \subset \widetilde{S}$ (with $m(N) = 0$ and $\widetilde{m}(\widetilde{N}) = 0$) such that for every $t \in T$, $s \in S \setminus N$ and $\widetilde{s} \in \widetilde{S} \setminus \widetilde{N}$, there is $t_n \in T_0$ satisfying $\lim f_{t_n}(s) = f_t(s)$ and $\lim \widetilde{f}_{t_n}(\widetilde{s}) = \widetilde{f}_t(\widetilde{s})$.

Propositions 2.3–2.5 are known collectively as *rigidity properties* of integral representations of $S\alpha S$ processes. As illustrated in the following examples, these are very stringent conditions.

Example 2.5. For $\alpha \in (0,2)$, $H \in (0,1)$, $H \neq 1/\alpha$ and $a, b \in \mathbb{R}$, the process

$$\{X_{a,b}(t)\}_{t\in\mathbb{R}} \stackrel{d}{=} \Big\{ \int_{\mathbb{R}} f_t(u) M(du) \Big\}_{t\in\mathbb{R}}, \tag{2.57}$$

where $M(du)$ is a $S\alpha S$ random measure on $\mathbb{R}$ with the Lebesgue control measure du, and

$$f_t(u) = a\big((t-u)_+^{H-1/\alpha} - (-u)_+^{H-1/\alpha}\big) + b\big((t-u)_-^{H-1/\alpha} - (-u)_-^{H-1/\alpha}\big), \tag{2.58}$$

is called a *linear fractional stable motion (LFSM)*. Here and throughout the work, $x_+ = \max\{x, 0\}$ is the positive part and $x_- = \max\{-x, 0\}$ is the negative part of $x \in \mathbb{R}$. It can be checked to be well defined, H-self-similar, and have stationary increments (e.g., Samorodnitsky and Taqqu [56] and Pipiras and Taqqu [42]). We will show that $X_{a,0}$ and $X_{0,b}$ have different laws. If $X_{a,0}$ and $X_{0,b}$ have the same law, then

$$\Big\{\int_{\mathbb{R}} a\big((t+u)_+^{H-1/\alpha} - u_+^{H-1/\alpha}\big)M(du)\Big\} \stackrel{d}{=} \Big\{\int_{\mathbb{R}} b\big((t+u)_-^{H-1/\alpha} - u_-^{H-1/\alpha}\big)M(du)\Big\}.$$

By the rigidity property (2.51), there are $\phi : \mathbb{R} \to \mathbb{R}$, $h : \mathbb{R} \to \mathbb{R}$ such that

$$a\big((t+u)_+^{H-1/\alpha} - u_+^{H-1/\alpha}\big) = bh(u)\big((t+\phi(u))_-^{H-1/\alpha} - (\phi(u))_-^{H-1/\alpha}\big) \quad \text{a.e. } dtdu.$$

Fix $u = u_0 < 0$ for which this relation holds a.e. dt. Then, for sufficiently large t, $t + u_0 > 0$ and $t + \phi(u_0) > 0$, so that

$$a(t+u_0)^{H-1/\alpha} = -bh(u_0)\phi(u_0)^{H-1/\alpha} \quad \text{a.e. } dt.$$

This is impossible since the right-hand side is just constant.

Remark 2.9. The findings in Example 2.5 stand in sharp contrast to the Gaussian case $\alpha = 2$. When $\alpha = 2$ and M is a Gaussian random measure, the process (2.57)–(2.58) is fractional Brownian motion with self-similarity parameter H (see Section 1.1) for any choice of a and b. That is, possibly up to a multiplicative constant (which depends on a and b), it is exactly the same process for fixed H. Similar observations can also be made in connection to the functions used in Remark 2.4. Consider the processes

$$X_\alpha(t) = \int_{\mathbb{R}} 1_{[0,t]}(u)M(du), \quad t \in \mathbb{R}, \tag{2.59}$$

and

$$\begin{aligned}\widehat{X}_\alpha(t) = &\int_{\mathbb{R}} \frac{1}{\sqrt{2\pi}} \frac{\sin(tx)}{x} |x|^{1-2/\alpha} M_1(du) \\ &+ \int_{\mathbb{R}} \frac{1}{\sqrt{2\pi}} \frac{(1-\cos(tx))}{x} |x|^{1-2/\alpha} M_2(du), \quad t \in \mathbb{R},\end{aligned} \tag{2.60}$$

where $M(du)$, $M_1(du)$, and $M_2(du)$ are independent $S\alpha S$ random measures on $\mathbb{R}$ with control measure du. The presence of $|x|^{1-2/\alpha}$ in the integrands of (2.60) is to make sure that the process $\widehat{X}_\alpha$ is well defined. In fact, the process $\widehat{X}_\alpha$ is known as the *real harmonizable fractional stable motion* with self-similarity parameter $H = 1/\alpha$ (see Section 7.7 in Samorodnitsky and Taqqu [56] and, in particular, Remark 1 in that section). The process X_α in (2.59), on the other hand, is just a $S\alpha S$ Lévy motion. It has independent increments and is self-similar with self-similarity parameter $H = 1/\alpha$. In the Gaussian case $\alpha = 2$, (2.59) and (2.60) are two different representations of the same process, which is Brownian motion (the representation (2.59) is in the time domain and the representation (2.60) is in the spectral domain). But in the $S\alpha S$ case $0 < \alpha < 2$, the rigidity results yield that these are two different $S\alpha S$ processes, which can be established by arguing as in (2.28).

Example 2.6. Consider two $S\alpha S$ linear time series

$$X_\alpha(n) = \sum_{k=-\infty}^{n} a_{n-k}\varepsilon_k, \quad Y_\alpha(n) = \sum_{k=-\infty}^{\infty} b_{n-k}\varepsilon_k, \quad n \in \mathbb{Z}, \tag{2.61}$$

where $\{a_n\}, \{b_n\} \in l^\alpha(\mathbb{Z})$ and $\{\varepsilon_k\}$ consists of i.i.d. $S\alpha S$ random variables such that

$$\mathbb{E}\exp\{i\theta\varepsilon_k\} = \exp\{-|\theta|^\alpha\}, \quad \theta \in \mathbb{R}.$$

Suppose $b_n \neq 0$ for infinitely many $n = -1, -2, \ldots$, and $a_n \neq 0$ for infinitely many $n = 0, 1, 2, \ldots$. That is, X_α can be thought as a causal time series, and Y_α as a noncausal time series. The rigidity property in Proposition 2.5 can be used to show that X_α and Y_α cannot have the same distribution. Suppose they do. Then, by Proposition 2.5, there are functions $h : \mathbb{Z} \to \mathbb{R}$ and $\phi : \mathbb{Z} \to \mathbb{Z}$ such that

$$a_{n-k} = h(k)\, b_{n-\phi(k)}$$

for all n and k (the "a.e." condition becomes "everywhere" since the spaces $T = \mathbb{Z}$, $S = \mathbb{Z}$, and $\widetilde{S} = \mathbb{Z}$ are discrete). Then, $h(k) \neq 0$ for any k since otherwise $a_{n-k} = 0$ for all n implies $a_\ell = 0$ for $\ell \geq 0$ (leading to contradiction). But if $h(k) \neq 0$, then $a_{n-k} = 0$ for $n - k < 0$ ($n < k$) implies that $b_{n-\phi(k)} = 0$ for $n < k$ and hence, in particular, $n < \min\{k, \phi(k)\}$. That is, $b_\ell = 0$ if $\ell < \ell_0$ for some ℓ_0, leading to a contradiction.

Remark 2.10. As with Example 2.5 and Remark 2.9, Example 2.6 also stands in sharp contrast to the Gaussian case $\alpha = 2$. In the Gaussian case $\alpha = 2$, when $\{\varepsilon_k\}$ consists of i.i.d. standard normal random variables, the noncausal (two-sided) time series Y_α in (2.61) can be represented as causal (one-sided) time series X_α in (2.61) under mild assumption on the so-called spectral density $f(\lambda)$ of Y_α, namely, $\int_{-\pi}^{\pi} \ln f(\lambda) d\lambda > -\infty$ (see, e.g., Brockwell and Davis [6], Remark 1 in Section 5.8 and Theorem 5.7.2).

Example 2.7. Suppose that a $S\alpha S$, $0 < \alpha < 2$, stationary process $X_\alpha(t)$, $t \in \mathbb{R}$, has two moving average representations, namely,

$$\{X_\alpha(t)\}_{t\in\mathbb{R}} \overset{d}{=} \Big\{ \int_{\mathbb{R}} f(t+u)M(du) \Big\}_{t\in\mathbb{R}} \overset{d}{=} \Big\{ \int_{\mathbb{R}} g(t+u)M(du) \Big\}_{t\in\mathbb{R}}, \tag{2.62}$$

where M is a $S\alpha S$ random measure on $\mathbb{R}$ with control measure du and $f, g \in L^\alpha(\mathbb{R}, du)$. Then, by Proposition 2.5, there are functions $h : \mathbb{R} \to \mathbb{R} \setminus \{0\}$ and $\phi : \mathbb{R} \to \mathbb{R}$ such that

$$f(t+u) = h(u)g(t+\phi(u)) \quad \text{a.e. } dtdu.$$

By making the change of variables $v = t + u$, $u = u$, we deduce that

$$f(v) = h(u)g(v - u + \phi(u)) \quad \text{a.e. } dvdu.$$

By Lemma 1.1, (ii), there is $u = u_0$ such that

$$f(v) = h(u_0)g(v - u_0 + \phi(u_0)) \quad \text{a.e. } dv,$$

that is,

$$f(v) = h_0 g(v + \tau_0) \quad \text{a.e. } dv, \tag{2.63}$$

for some $h_0 \neq 0$, $\tau_0 \in \mathbb{R}$. In fact, since the scale coefficients $\|f\|_{L^\alpha(\mathbb{R},du)}$ and $\|g\|_{L^\alpha(\mathbb{R},du)}$ of $X_\alpha(0)$ are equal, we deduce that $|h_0| = 1$, that is, $h_0 = \pm 1$. The result (2.63) was established originally by Kanter [22]. It is also stated as Lemma 5.4 of Samorodnitsky and Taqqu [56]. We shall formulate it as a corollary.

Corollary 2.1. *Suppose that (2.62) holds. Then, there are $\varepsilon \in \{-1, 1\}$ and $\tau \in \mathbb{R}$ such that*

$$f(t) = \varepsilon g(t + \tau) \quad \text{a.e. } dt.$$

This corollary can be used to verify the following example, which generalizes Example 2.5.

Example 2.8. Let $\{X_{a,b}(t)\}_{t\in\mathbb{R}}$ be the LFSM defined by (2.57) and suppose $c_{a,b}$ is such that the scale parameter of the $S\alpha S$ random variable $X_{a,b}(1)/c_{a,b}$ is 1. Then, by using Corollary 2.1, one can show that, with $|a| + |b| > 0$ and $|a'| + |b'| > 0$,

$$\left\{\frac{1}{c_{a,b}} X_{a,b}(t)\right\}_{t\in\mathbb{R}} \stackrel{d}{=} \left\{\frac{1}{c_{a',b'}} X_{a',b'}(t)\right\}_{t\in\mathbb{R}}$$

if and only if one of the following conditions holds:

(i) $a = a' = 0$,
(ii) $b = b' = 0$,
(iii) a, a', b, b' are nonzero and $a/b = a'/b'$.

See Samorodnitsky and Taqqu [56], Theorem 7.4.5, for more details.

We have used Proposition 2.5 which does not require that the representations (2.18) be minimal. That proposition, therefore, is quite handy and can be used to show, as in Example 2.5, that two given representations are different. Note that (2.51) holds not only "a.e. $\widetilde{m}(d\widetilde{s})$" but also "a.e. $\sigma(dt)$," which explains the presence of "a.e. $dtdu$" in Example 2.5.

In the following sections, we will also apply the other rigidity properties which require that minimality holds.

2.3 Nonsingular Flows and Their Functionals

The rigidity properties presented in Section 2.2 have fundamental implications for $S\alpha S$ processes with an invariance property.

2.3.1 Rigidity and Flows

Suppose, for example, that a $S\alpha S$ process $X_\alpha = \{X_\alpha(t)\}_{t\in\mathbb{R}}$ is stationary and expressed through a minimal representation

$$\{X_\alpha(t)\}_{t\in\mathbb{R}} \stackrel{d}{=} \Big\{ \int_S f_t(s) M(ds) \Big\}_{t\in\mathbb{R}},$$

where $\{f_t\}_{t\in\mathbb{R}} \subset L^\alpha(S,\mathscr{B}(S),m)$ and a $S\alpha S$ random measure M on S has control measure m. The stationarity of X means that, for any $h \in \mathbb{R}$,

$$\Big\{ \int_S f_t(s) M(ds) \Big\}_{t\in\mathbb{R}} \stackrel{d}{=} \Big\{ \int_S f_{t+h}(s) M(ds) \Big\}_{t\in\mathbb{R}}.$$

The representations on both sides are minimal. We can therefore apply Proposition 2.4 and conclude that there are

$$\phi_h : S \to S \quad \text{and} \quad \varepsilon_h : S \to \{-1,1\}$$

such that, for every $t,h \in \mathbb{R}$,

$$f_{t+h}(s) = \varepsilon_h(s) \Big\{ \frac{d(m\circ\phi_h)}{dm}(s) \Big\}^{1/\alpha} f_t(\phi_h(s)) \quad \text{a.e. } m(ds). \tag{2.64}$$

Writing this with $h = h_1 + h_2$ leads to

$$f_{t+h_1+h_2}(s) = \varepsilon_{h_1+h_2}(s) \Big\{ \frac{d(m\circ\phi_{h_1+h_2})}{dm}(s) \Big\}^{1/\alpha} f_t(\phi_{h_1+h_2}(s)) \quad \text{a.e. } m(ds). \tag{2.65}$$

On the other hand, by applying (2.64) in *two steps*, we get that

$$f_{t+h_1+h_2}(s) = \varepsilon_{h_1}(s) \Big\{ \frac{d(m\circ\phi_{h_1})}{dm}(s) \Big\}^{1/\alpha} f_{t+h_2}(\phi_{h_1}(s)) = \varepsilon_{h_1}(s)\varepsilon_{h_2}(\phi_{h_1}(s)) \times$$

$$\times \Big\{ \frac{d(m\circ\phi_{h_1})}{dm}(s) \Big\}^{1/\alpha} \Big\{ \frac{d(m\circ\phi_{h_2})}{dm}(\phi_{h_1}(s)) \Big\}^{1/\alpha} f_t(\phi_{h_2}(\phi_{h_1}(s))) \quad \text{a.e. } m(ds). \tag{2.66}$$

In view of (2.65) and (2.66), the uniqueness now yields that, for any $h_1, h_2 \in \mathbb{R}$,

$$\phi_{h_1+h_2}(s) = \phi_{h_2}(\phi_{h_1}(s)), \quad \varepsilon_{h_1+h_2}(s) = \varepsilon_{h_1}(s)\varepsilon_{h_2}(\phi_{h_1}(s)) \quad \text{a.e. } m(ds). \tag{2.67}$$

The first relation in (2.67) defines the so-called "flows" which are discussed in greater detail in Section 2.3.2. Flows arise not only in the stationary context as above but also for other invariance properties such as self-similarity and stationarity of increments. The second relation in (2.67) defines the so-called cocycle associated with a flow $\{\phi_h\}$. Cocycles and other functionals of flows are discussed in Section 2.3.4.

2.3.2 Flows

We begin with the definition of a flow. As in previous sections, we suppose that $(S, \mathscr{B}(S), m)$ is a standard Lebesgue space.

Definition 2.2. A collection of maps $\{\phi_h\}_{h\in\mathbb{R}}$, $\phi_h : S \to S$, satisfying

$$\phi_{h_1+h_2}(s) = \phi_{h_2}(\phi_{h_1}(s)), \quad \phi_0(s) = s, \quad s \in S,\ h_1, h_2 \in \mathbb{R}, \tag{2.68}$$

is called a (additive) *flow*. A *multiplicative flow* is a collection of maps $\{\psi_c\}_{c>0}$, $\psi_c : S \to S$, satisfying

$$\psi_{c_1c_2}(s) = \psi_{c_2}(\psi_{c_1}(s)), \quad \psi_1(s) = s, \quad s \in S,\ c_1, c_2 > 0. \tag{2.69}$$

In view of (2.68), one can think of $\phi_h(s)$ as the position of a moving particle s at time h. The position $\phi_{h_1+h_2}(s)$ of the particle at time $h_1 + h_2$ can be obtained from the position $\phi_{h_1}(s)$ at time h_1 by running the flow more h_2 units of time to the position $\phi_{h_2}(\phi_{h_1}(s))$.

We use throughout the notation $\{\phi_h\}_{h\in\mathbb{R}}$ for an additive flow, and the notation $\{\psi_c\}_{c>0}$ for a multiplicative flow. Note that $\{\phi_h\}_{h\in\mathbb{R}}$ is a (additive) flow if and only if $\{\psi_c = \phi_{\ln c}\}_{c>0}$ is a multiplicative flow. In the context of self-similar processes below, we shall work with multiplicative flows. In this section, however, the results will be presented for additive flows only. These can be translated easily to multiplicative flows by using the aforementioned relation, and will be used for multiplicative flows without repeating this connection.

The relation (2.68) implies that each ϕ_h is one-to-one and onto, and has a measurable inverse $\phi_h^{-1} = \phi_{-h}$. Indeed, this follows by taking $h_1 = h$ and $h_2 = -h$ in (2.68) to get $\phi_h(\phi_{-h}(s)) = \phi_{-h}(\phi_h(s)) = s$. We shall work with flows which are *nonsingular*, that is, for any $h \in \mathbb{R}$ and $A \in \mathscr{B}(S)$,

$$m(A) = 0 \quad \Leftrightarrow \quad m(\phi_h^{-1}(A)) = 0. \tag{2.70}$$

This condition essentially states that no mass disappears or reappears under the flow. The flows considered throughout will also be *measurable*, that is, the map $\phi_h(s) : \mathbb{R} \times S \to S$ is measurable.

The following provides a few basic examples of flows.

Example 2.9. The collection of maps

$$\phi_h(s) = s, \quad s \in S,\ h \in \mathbb{R}, \tag{2.71}$$

defines a flow. It is a *fixed (identity)* flow[12] since particles do not move under this flow.

Example 2.10. Consider $(S, \mathscr{B}(S), m) = (\mathbb{R}, \mathscr{B}(\mathbb{R}), \text{Lebesgue})$. The collection of maps

[12] The terms "fixed" and "identity" are used interchangeably and labeled by the letter F.

$$\phi_h(u) = u + h, \quad u \in \mathbb{R},\ h \in \mathbb{R}, \tag{2.72}$$

defines a flow. Under this flow, particles move to the right of the horizontal axis at a constant speed 1. The flow is nonsingular. The corresponding multiplicative flow is

$$\psi_c(u) = u + \ln c, \quad u \in \mathbb{R},\ c > 0,$$

so that

$$\psi_{c_2}(\psi_{c_1}(u)) = \psi_{c_2}(u + \ln c_1) = u + \ln c_1 + \ln c_2 = u + \ln c_1 c_2 = \psi_{c_1 c_2}(u).$$

Example 2.11. Consider $(S, \mathscr{B}(S), m) = ([0,1), \mathscr{B}([0,1)), \text{Lebesgue})$ and let

$$\{u\}_1 = u - [u]_1, \quad u \in \mathbb{R}, \tag{2.73}$$

be the fractional part of u, where $[u]_1$ is the integer part of u (the largest integer which is less than or equal to u). For example, $\{2\}_1 = 0$, $\{2.3\}_1 = 0.3$, $\{-2.3\}_1 = -2.3 - (-3) = 0.7$, etc. The collection of maps

$$\phi_h(u) = \{u + h\}_1, \quad u \in [0,1),\ h \in \mathbb{R}, \tag{2.74}$$

defines a flow. This can be seen from

$$\begin{aligned}
\phi_{h_1+h_2}(u) &= \{u + h_1 + h_2\}_1 = u + h_1 + h_2 - [u + h_1 + h_2]_1 \\
&= u + h_1 + h_2 - [[u + h_1]_1 + \{u + h_1\}_1 + h_2]_1 \\
&= u + h_1 + h_2 - [u + h_1]_1 - [\{u + h_1\}_1 + h_2]_1 && (2.75) \\
&= \{u + h_1\}_1 + h_2 - [\{u + h_1\}_1 + h_2]_1 = \{\{u + h_1\}_1 + h_2\}_1 && (2.76) \\
&= \phi_{h_2}(\phi_1(u)),
\end{aligned}$$

where in (2.75), we used the fact that $[n + a]_1 = n + [a]_1$ for any $n \in \mathbb{Z}$ and $a \in \mathbb{R}$, and in (2.76), we used the definition (2.73) twice. Under the flow (2.74), a particle moves to the right of the interval $[0,1)$ at a constant speed 1, jumping back to 0 once it reaches 1. The flow can be viewed as occurring on a circle (after identifying the endpoints of $[0,1)$), and will be called *cyclic*.

2.3.3 Hopf Decomposition of a Flow

$S\alpha S$ processes will be decomposed based on the structure of the underlying flows. The so-called Hopf decomposition is one such fundamental structural result for flows (see, for example, Krengel [26], p. 17, and Rosiński [46], p. 1171). Consider a nonsingular map $V : S \to S$. A set $B \in \mathscr{B}(S)$ is called *wandering* if the sets

$$V^{-k}B = \{s \in S : V^k(s) \in B\}, \quad k \geq 1,$$

are disjoint. The map V is called *conservative* if there is no wandering set of positive measure m. Given any nonsingular map $V : S \to S$, there exists a decomposition of S into two disjoint measurable sets C and D such that

(i) C and D are V-invariant, that is, $V^{-1}C = C$ and $V^{-1}D = D$,
(ii) the restriction of V to C is conservative, and
(iii) $D = \cup_{k=-\infty}^{\infty} V^k B$ for some wandering set B.

The decomposition of S into sets D and C is unique (modulo m). It is called the *Hopf decomposition*. The sets D and C are called the *dissipative part* and the *conservative part*, respectively.

If $\{\phi_h\}_{h\in\mathbb{R}}$ is a nonsingular flow, then for every $h \in \mathbb{R}$, each nonsingular map ϕ_h has the Hopf decomposition

$$S = D_h \cup C_h.$$

One can show (Krengel [25], Lemma 2.7) that all D_h, $h \neq 0$, are equal to each other modulo m and that there is a set D, invariant under the flow, such that for every $h \neq 0$, $D = D_h$ modulo m. The decomposition of the space S into the sets D and $C := S \setminus D$, that is,

$$S = D + C, \tag{2.77}$$

is called the *Hopf decomposition for the flow* $\{\phi_h\}_{h\in\mathbb{R}}$. A flow is called *dissipative* if $S = D$ and *conservative* if $S = C$ (modulo m). We also have the same notions for a nonsingular multiplicative flow $\{\psi_c\}_{c>0}$.

A useful representation for the dissipative and the conservative parts of an additive flow $\{\phi_h\}_{h\in\mathbb{R}}$ is given by:

Proposition 2.6.

$$D = \left\{ s \in S : \int_{\mathbb{R}} g(\phi_h(s)) \frac{d(m\circ\phi_h)}{dm}(s) dh < \infty \right\} \quad a.e.\ dm, \tag{2.78}$$

$$C = \left\{ s \in S : \int_{\mathbb{R}} g(\phi_h(s)) \frac{d(m\circ\phi_h)}{dm}(s) dh = \infty \right\} \quad a.e.\ dm, \tag{2.79}$$

where g is any $L^1(S,m)$ function such that $g \geq 0$ a.e. and $\mathrm{supp}\{g\} = S$ *a.e.*

This fact follows from the proof of Theorem 4.1 in Rosiński [46] (see also Lemma 2.7 in Krengel [25]). In fact, a slightly stronger statement is the following (see (2.4) and (2.5) in Rosiński [46]): if D_0 and C_0 denote the right-hand sides of (2.78) and (2.79), then

$$D \subset D_0, \quad C \cap \mathrm{supp}\{g\} \subset C_0 \quad \text{a.e. } dm. \tag{2.80}$$

(If, in particular, as we suppose in Proposition 2.6, $\mathrm{supp}\{g\} = S$ a.e., then we recover (2.78) and (2.79). Indeed, (2.80) becomes $D \subset D_0$, $C \subset C_0$ a.e. and since $S = D + C = D_0 + C_0$, this yields $D = D_0$, $C = C_0$ a.e.) Note that the choice of the function g in Proposition 2.6 is arbitrary as long as $g \geq 0$. We also assume

$\text{supp}\{g\} = S$ a.e. to get good coverage. The function g plays the role of a test function. If $\text{supp}\{g\}$ is smaller than S, then we obtain the slightly stronger statement given in (2.80).

Example 2.12. Consider the flows in Examples 2.9–2.11. The fixed flow (2.71) is clearly conservative. Indeed, since m is Lebesgue measure on $\mathbb{R}$, we have

$$\frac{d(m \circ \phi_h)}{dm}(s) = \frac{d\phi_h(s)}{ds} = \frac{ds}{ds} = 1$$

and thus

$$\int_{\mathbb{R}} g(\phi_h(s))\frac{d(m \circ \phi_h)}{dm}(s)dh = \int_{\mathbb{R}} g(s)dh = g(s)\int_{\mathbb{R}} dh = \infty.$$

The cyclic flow (2.74) is also conservative, since the integral in (2.78)–(2.79) is

$$\int_{\mathbb{R}} g(\phi_h(s))\frac{d(m \circ \phi_h)}{dm}(s)dh = \int_{\mathbb{R}} g(\{u+h\}_1)dh = \int_{\mathbb{R}} g(\{h\}_1)dh = \infty.$$

Similarly, the flow (2.72) is dissipative since

$$\int_{\mathbb{R}} g(\phi_h(s))\frac{d(m \circ \phi_h)}{dm}(s)dh = \int_{\mathbb{R}} g(u+h)dh = \int_{\mathbb{R}} g(h)dh < \infty.$$

The next result, due to Krengel [25], p. 19 (see also Rosiński [46], p. 1176), characterizes dissipative flows. It shows that a dissipative flow is null-isomorphic to a mixture of flows (2.72). The notion of null-isomorphism is described following the theorem.

Theorem 2.3. *(Krengel) Every measurable nonsingular dissipative flow $\{\phi_h\}_{h\in\mathbb{R}}$ on a standard Lebesgue space is null-isomorphic* (mod 0) *to a flow $\{\widetilde{\phi}_h\}_{h\in\mathbb{R}}$ on some standard Lebesgue space $(Y \times \mathbb{R}, \mathscr{B}(Y \times B), \nu(dy)du)$ defined by*

$$\widetilde{\phi}_h(y,u) = (y, u+h), \quad (y,u) \in Y \times \mathbb{R},\ h \in \mathbb{R}. \tag{2.81}$$

In the case of a multiplicative flow $\{\psi_c\}_{c>0}$, replace $\widetilde{\phi}_h$ by

$$\widetilde{\psi}_c(y,u) = (y, u+\ln c), \quad (y,u) \in Y \times \mathbb{R},\ c > 0. \tag{2.82}$$

Null-isomorphism (mod 0) in Theorem 2.3 means that there exist two null sets $N \subset S$ and $N' \subset Y \times \mathbb{R}$, and a null-isomorphism

$$\Phi : Y \times \mathbb{R} \setminus N' \to S \setminus N$$

such that

$$\phi_h(\Phi(y,u)) = \Phi(\widetilde{\phi}_h(y,u)),$$

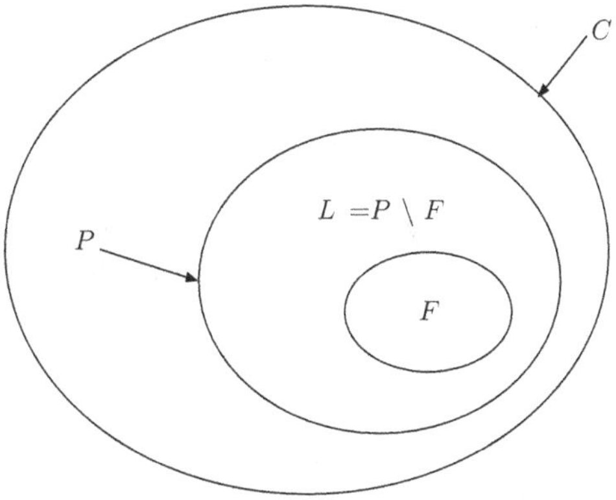

Fig. 2.1 The sets C, P, L, and F.

for all $c > 0$ and $(y,u) \in Y \times \mathbb{R} \setminus N'$. The sets $S \setminus N$ and $Y \times \mathbb{R} \setminus N'$ above are invariant under the flows $\{\phi_h\}_{h\in\mathbb{R}}$ and $\{\widetilde{\phi}_h\}_{h\in\mathbb{R}}$, respectively, i.e., $\phi_h(S\setminus N) = S\setminus N$ and similarly for $\widetilde{\phi}_h$.

Theorem 2.3 offers a nice characterization of dissipative flows. If the flow in the space S is dissipative, one can add a dimension. There is a space Y such that the flow on S can be described by a new flow on $Y \times \mathbb{R}$. This new flow has the following interesting characteristics described in (2.81). It stays put on Y and it has on $\mathbb{R}$ the dissipative motion $u \to u+h$, $h \in \mathbb{R}$, described in Example 2.10.

There is no such simple characterization of conservative flows. In the context of stable processes, we will nevertheless use several classes of conservative flows. Given a general flow, these classes will be associated with certain subsets of the conservative part C of the flow. More specifically, let $\{\phi_h\}_{h\in\mathbb{R}}$ be a measurable flow on a standard Lebesgue space $(S, \mathscr{B}(S), m)$. We shall use the sets P, F, and L, defined below, which are subsets of C. See also Figure 2.1.

Definition 2.3. Let

$$P := \{s : \exists\, p = p(s) \in \mathbb{R}\setminus\{0\} : \phi_p(s) = s\}, \tag{2.83}$$
$$F := \{s : \phi_h(s) = s \text{ for all } h \in \mathbb{R}\}, \tag{2.84}$$
$$L := P\setminus F \tag{2.85}$$

be the *periodic*, *fixed (identity)*, and *cyclic* points of the flow $\{\phi_h\}_{h\in\mathbb{R}}$, respectively.[13]

Definition 2.4. A measurable flow $\{\phi_h\}_{h\in R}$ on $(S, \mathscr{B}(S), m)$ is *periodic* if $S = P$ m-a.e., is *fixed (identity)* if $S = F$ m-a.e., and it is *cyclic* if $S = L$ m-a.e.

[13]We use F for fixed or identity. We use D for dissipative and C for conservative, so we cannot use C for cyclic. We use L instead to refer to cycLic.

Informally, the *periodic* flows are conservative flows, such that, through them, each point of the space comes back to its initial position with a finite return time, positive or null (where the return time for a point is defined as the infimum of positive times that the flow comes back to the point). The flows are *fixed* or *identity* if the return time is null and the flow is *cyclic* if the return time is positive. We will see in Section 3.2.8 that not all conservative flows are periodic.

The following auxiliary lemma concerns the measurability of the introduced sets.

Lemma 2.3. *The set F in (2.84) is (Borel) measurable and the sets P in (2.83) and L in (2.85) are m-measurable.*

PROOF: To show that the set F of fixed points is measurable, view $\phi_h(s)$ as a function of h and s. If one shows that

$$F' := \{s : \phi_h(s) = s \text{ a.e. } dh\}$$

is measurable and $F' = F$, then it will follow that F is measurable. We first show that $F' = F$. Note that the inclusion $F \subset F'$ follows immediately from the definitions of F and F'. We thus only need to show that $F' \subset F$. If $s \in F'$, then $\phi_h(s) = s$ a.e. dh, that is,

$$\tau := \{h : \phi_h(s) = s\} = \mathbb{R} \quad \text{a.e. } dh.$$

Observe that τ is an additive group of $\mathbb{R}$ (if $h_1, h_2 \in \tau$, then $h_1 + h_2 \in \tau$ because $\phi_{h_1+h_2}(s) = \phi_{h_1}(\phi_{h_2}(s)) = \phi_{h_1}(s) = s$) and hence by Corollary 1.1.4 in Bingham et al. [5], we have $\tau = \mathbb{R}$, that is, $\phi_h(s) = s$ for all $h \in \mathbb{R}$. This shows that $F' \subset F$, and hence $F' = F$.

We now show that F' is measurable. Note that, in contrast to F, F' is defined by requiring $\phi_h(s) = s$ only a.e. dh. Then, F' and hence F can be expressed as

$$F = F' = \{s : k(s) = 0\},$$

where

$$k(s) = \int_{\mathbb{R}} 1_{\{\phi_h(s) \neq s\}}(h,s)dh.$$

(Indeed, $k(s) = 0$ is equivalent to $1_{\{\phi_h(s) \neq s\}}(h,s) = 0$ a.e. dh, or to $\phi_h(s) = s$ a.e. dh.) Since the function $k(\cdot)$ is measurable by Fubini's theorem, the set F is measurable as well (use Theorem A in Halmos [18], p. 143).

To prove that the set

$$P = \{s : \exists\, p = p(s) \neq 0 : \phi_p(s) = s\}$$

is m-measurable, consider the measurable set

$$\widetilde{P} = \{(s,p) : \phi_p(s) = s, p \neq 0\}.$$

Observe that

$$P = \text{proj}_S\{\widetilde{P}\} := \{s : \exists\, p : (s,p) \in \widetilde{P}\}.$$

The m-measurability of P follows from Lemma B.1 in Appendix B. The set L is m-measurable because $L = P \setminus F$. $\square$

We shall use in the sequel the following alternative characterization of cyclic flows. For $a > 0$ and $x \in \mathbb{R}$,

$$[x]_a = \max\{n \in \mathbb{Z} : na \le x\}, \quad \{x\}_a = x - a[x]_a \ge 0. \tag{2.86}$$

Proposition 2.7. *A measurable flow $\{\phi_h\}_{h\in\mathbb{R}}$ on $(S, \mathscr{B}(S), m)$ is cyclic if and only if it is null-isomorphic (mod 0) to a flow*

$$\widetilde{\phi}_h(z, v) = (z, \{v+h\}_{q(z)}) \tag{2.87}$$

on

$$(Z \times [0, q(\cdot)), \mathscr{B}(Z \times [0, q(\cdot))), \sigma(dz)\lambda(dv)),$$

where $q(z) > 0$ a.e. is some measurable function.

It is interesting to compare Proposition 2.7 with Theorem 2.3 which is valid for dissipative flows.

The proposition is proved in Pipiras and Taqqu [38], Theorem 2.1. The sufficiency part of the proposition is elementary to prove. The necessity part uses the so-called special flow $\{\widetilde{\phi}_t\}_{t\in T}$, which we introduce for reference. Informally, the flow $\widetilde{\phi}_t(y, u)$ is defined on the set of points

$$\Omega = \{(y, u) : 0 \le u < r(y), y \in Y\} = Y \times [0, r(\cdot)),$$

where $r(y)$ is a positive function. Plotting (y, u) in two dimensions, we can view the flow $\widetilde{\phi}_t$ as moving up vertically at constant speed till it reaches the level $r(y)$, and then jumps back to a point $(y', 0)$ before it renews its vertical climb, this time from the point y' (see Figure 2.2). Thus, if one focuses only on the horizontal Y axis, the flow starting at y moves to $y' = Vy$, then to $V^2 y, \ldots, V^n y, \ldots$ Since the flow $\widetilde{\phi}_t$ moves constantly, observe that it has no fixed points.

The flow $\widetilde{\phi}_t$ is defined formally as follows. Let $(Y, \mathscr{B}(Y), \tau)$ be a standard Lebesgue space, V be a null-isomorphism of Y onto itself, and r be a positive measurable function on Y such that

$$\sum_{k=0}^{\infty} r(V^k y) = \sum_{k=-\infty}^{-1} r(V^k y) = \infty.$$

Set

$$\Omega = \{(y, u) : 0 \le u < r(y), y \in Y\}, \quad \mathscr{E} = \mathscr{B}(Y \times [0, r(\cdot)))$$

and P be a measure on $(\Omega, \mathscr{E})$ such that

$$dP(y, u) = p(y, u)\tau(dy)du \quad \text{and} \quad P(\Omega) = 1.$$

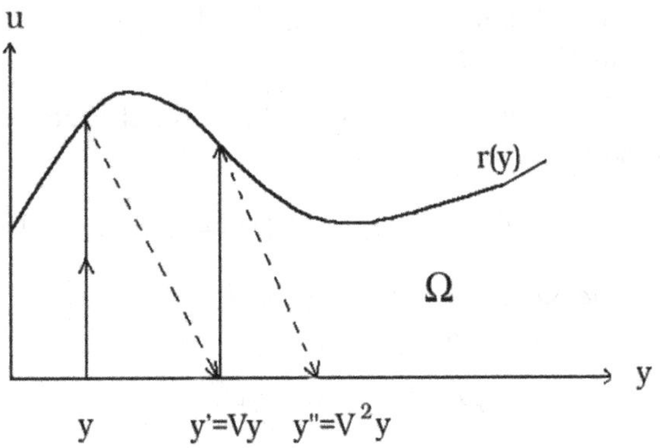

Fig. 2.2 View the flow $\widetilde{\phi}_t$ as moving up vertically at constant speed till it reaches the level $r(y)$, and then jumps back to a point $(y',0)$ before it renews its vertical climb, this time from the point y'.

Consider now the map defined on Ω by

$$\widetilde{\phi}_h(y,u) = (V^n y, u+h-r_n(y)) \tag{2.88}$$

for

$$0 \le u+h-r_n(y) < r(V^n y),$$

where

$$r_n(y) = \begin{cases} \sum_{k=0}^{n-1} r(V^k y), & \text{if } n \ge 1, \\ 0, & \text{if } n = 0, \\ \sum_{k=n}^{-1} r(V^k y), & \text{if } n \le -1. \end{cases}$$

One can verify that $\{\widetilde{\phi}_h\}_{h\in\mathbb{R}}$ is a (measurable, nonsingular) flow on $(\Omega,\mathscr{E},P)$. It is called a *special flow* built under the function r. According to Theorem 3.1 in Kubo [27], a (measurable, nonsingular) flow $\{\phi_h\}_{h\in\mathbb{R}}$ without fixed points on a standard Lebesgue space is null-isomorphic (mod 0) to some special flow $\{\widetilde{\phi}_h\}_{h\in\mathbb{R}}$ built under the function r.

Finally, note the following.

Remark 2.11. The sets P, F, and L are subsets of the conservative part C of the flow. In view of the Hopf decomposition (2.77), we can write (m-a.e.)

$$S = D+C = D+F+P\setminus F+C\setminus P = D+F+L+(C\setminus P). \tag{2.89}$$

Different classes of stable processes derived below will be associated with these different sets of underlying flows.

2.3.4 Cocycles

The notion of cocycle is fundamental.

Definition 2.5. Let $\{\phi_h\}_{h\in\mathbb{R}}$ be a flow on $(S,\mathscr{B}(S))$. A measurable map $\varepsilon_h(s):\mathbb{R}\times S\to\{-1,1\}$ satisfying

$$\varepsilon_{h_1+h_2}(s)=\varepsilon_{h_1}(s)\varepsilon_{h_2}(\phi_{h_1}(s)),\quad s\in S,\ h_1,h_2\in\mathbb{R}, \tag{2.90}$$

is called a *cocycle* associated with the flow ϕ. If (2.90) holds a.e. $m(ds)$ for each fixed $h_1,h_2\in\mathbb{R}$, then $\{\varepsilon_h\}_{h\in\mathbb{R}}$ is called an *almost cocycle*. For a multiplicative flow $\{\psi_c\}_{c>0}$, the cocycle equation is

$$\varepsilon_{c_1c_2}(s)=\varepsilon_{c_1}(s)\varepsilon_{c_2}(\psi_{c_1}(s)),\quad s\in S,\ c_1,c_2>0. \tag{2.91}$$

Remark 2.12. We shall also use below maps satisfying (2.90) but which take values in $\mathbb{R}$ (not necessarily in $\{-1,1\}$). We shall continue referring to these maps as cocycles. But to distinguish them from cocycles as defined above, we shall also indicate that their values fall in $\mathbb{R}$.

In the following three examples, we characterize cocycles associated with dissipative, fixed, and cyclic flows.

Example 2.13. If $\{\phi_h\}_{h\in\mathbb{R}}$ is a fixed flow (2.71), the cocycle equation (2.90) becomes

$$\varepsilon_{h_1+h_2}(s)=\varepsilon_{h_1}(s)\varepsilon_{h_2}(s),\quad s\in S,\ h_1,h_2\in\mathbb{R}.$$

In particular, $\varepsilon_h(s)=(\varepsilon_{h/2}(s))^2$. Since $\varepsilon_{h/2}(s)=1$ or -1, we deduce that $\varepsilon_h(s)=1$, $s\in S$, $h\in\mathbb{R}$.

Example 2.14. By Theorem 2.3, any dissipative flow is null-isomorphic to a dissipative flow $\{\widetilde{\phi}_h\}_{h\in\mathbb{R}}$ given in (2.81). If $\{\widetilde{\varepsilon}_h\}_{h\in\mathbb{R}}$ is a cocycle associated with the flow $\{\widetilde{\phi}_h\}_{h\in\mathbb{R}}$, the cocycle equation (2.90) becomes

$$\widetilde{\varepsilon}_{h_1+h_2}(y,u)=\widetilde{\varepsilon}_{h_1}(y,u)\widetilde{\varepsilon}_{h_2}(y,u+h_1),\quad (y,u)\in Y\times\mathbb{R},\ h_1,h_2\in\mathbb{R}.$$

Setting $u=0$, $h_2=h$, and $h_1=v$, we conclude that

$$\widetilde{\varepsilon}_h(y,v)=\frac{\widetilde{\varepsilon}_{v+h}(y,0)}{\widetilde{\varepsilon}_v(y,0)}=\frac{\widetilde{\varepsilon}(y,v+h)}{\widetilde{\varepsilon}(y,v)}$$

for a function $\widetilde{\varepsilon}:Y\times\mathbb{R}\to\{-1,1\}$. For a multiplicative dissipative flow, the cocycle can be expressed similarly as

$$\widetilde{b}_c(y,s)=\frac{\widetilde{b}(y,s+\ln c)}{\widetilde{b}(y,s)} \tag{2.92}$$

for a measurable function $\widetilde{b}:Y\times\mathbb{R}\to\{-1,1\}$.

Example 2.15. By Proposition 2.7, any cyclic flow is null-isomorphic to a cyclic flow $\{\widetilde{\phi}_h\}_{h\in\mathbb{R}}$ given in (2.87). If $\{\widetilde{\varepsilon}_h\}_{h\in\mathbb{R}}$ is a cocycle associated with the flow $\{\widetilde{\phi}_h\}_{h\in\mathbb{R}}$, we will show that

$$\widetilde{\varepsilon}_h(z,v) = \widetilde{\varepsilon}_0(z)^{[v+h]_{q(z)}} \frac{\widetilde{\varepsilon}(z,\{v+h\}_{q(z)})}{\widetilde{\varepsilon}(z,h)} \tag{2.93}$$

for functions $\widetilde{\varepsilon}_0 : Z \to \{-1,1\}$ $\widetilde{\varepsilon} : Z\times\mathbb{R} \to \{-1,1\}$. We consider the case $Z = \{1\}$ only and denote $\widetilde{\varepsilon}_h(z,v)$ by $\widetilde{\varepsilon}(h,v)$ and $q(z)$ by q. The proof for a general space Z follows as below by working with a fixed z. With the new notation and in the considered special case, the cocycle equation (2.90) reads

$$\widetilde{\varepsilon}(h_1+h_2,v) = \widetilde{\varepsilon}(h_1,v)\widetilde{\varepsilon}(h_2,\{v+h_1\}_q), \tag{2.94}$$

for any $h_1,h_2 \in \mathbb{R}$ and $v \in [0,q)$.

By taking $v = 0$ in (2.94), we get

$$\widetilde{\varepsilon}(h_1+h_2,0) = \widetilde{\varepsilon}(h_1,0)\widetilde{\varepsilon}(h_2,\{h_1\}_q).$$

Then, by setting $h_2 = h$ and $h_1 = v$,

$$\widetilde{\varepsilon}(h,v) = \frac{\widetilde{\varepsilon}(v+h,0)}{\widetilde{\varepsilon}(v,0)} \tag{2.95}$$

if $v \in [0,q)$. Observe now that, by (2.86) and (2.94),

$$\widetilde{\varepsilon}(v+h,0) = \widetilde{\varepsilon}(q[v+h]_q + \{v+h\}_q, 0) = \widetilde{\varepsilon}(q[v+h]_q,0)\widetilde{\varepsilon}(\{v+h\}_q,0)$$

for all $h \in \mathbb{R}$ and $v \in [0,q)$. Then, by (2.95),

$$\widetilde{\varepsilon}(h,v) = \frac{\widetilde{\varepsilon}(q[v+h]_q,0)\widetilde{\varepsilon}(\{v+h\}_q,0)}{\widetilde{\varepsilon}(v,0)} = \frac{\widetilde{\varepsilon}(q[v+h]_q)\widetilde{\varepsilon}(\{v+h\}_q)}{\widetilde{\varepsilon}(v)}, \tag{2.96}$$

where $\widetilde{\varepsilon}(\cdot) = \widetilde{\varepsilon}(\cdot,0)$. But by setting $h_1 = mq$, $h_2 = nq$, and $v = 0$ in (2.94), we get that $\widetilde{\varepsilon}(mq+nq) = \widetilde{\varepsilon}(mq)\widetilde{\varepsilon}(nq)$ for all $m,n \in \mathbb{Z}$. It follows that $\widetilde{\varepsilon}(qm) = \widetilde{\varepsilon}_0^m$ for some $\widetilde{\varepsilon} \in \{-1,1\}$ and hence, from (2.96), that

$$\widetilde{\varepsilon}(h,v) = \widetilde{\varepsilon}_0^{[v+h]_q} \frac{\widetilde{\varepsilon}(\{v+h\}_q)}{\widetilde{\varepsilon}(v)},$$

which proves (2.93) when $Z = \{1\}$.

For later reference, we include the result stating that an almost cocycle can be modified to a cocycle. A collection $\{\varepsilon'_h\}_{h\in\mathbb{R}}$ is a modification of $\{\varepsilon_h\}_{h\in\mathbb{R}}$ if $m(\varepsilon_h \neq \varepsilon_h) = 0$ for every $h \in \mathbb{R}$.

Theorem 2.4. *An almost cocycle* $\{\varepsilon_h\}_{h\in\mathbb{R}}$ *has a modification* $\{\varepsilon'_h\}_{h\in\mathbb{R}}$ *which is a cocycle.*

Theorem 2.4 is proved in a more general setting in Zimmer [68], Theorem B.9, p. 200. There, almost cocycles are called cocycles and cocycles are called strict cocycles. Another proof, based on the approach of Kubo [28] using special flows, can be found in Pipiras and Taqqu [35], Proposition 3.2.

2.3.5 Semi-Additive Functionals

We shall use below two other functionals related to a flow.

Definition 2.6. A *1-semi-additive functional* $\{g_h\}_{h\in\mathbb{R}}$ for a measurable flow $\{\phi_h\}_{h\in\mathbb{R}}$ is a measurable function $g_h : \mathbb{R}\times S \to \mathbb{R}$ satisfying

$$g_{h_1+h_2}(s) = e^{-h_2} g_{h_1}(s) + g_{h_2}(\phi_{h_1}(s)), \quad s\in S,\ h_1,h_2\in\mathbb{R}. \tag{2.97}$$

If (2.97) holds a.e. $m(ds)$ for each fixed $h_1,h_2\in\mathbb{R}$, the functional $\{g_h\}_{h\in\mathbb{R}}$ is called an *almost 1-semi-additive functional*. For a multiplicative flow $\{\psi_c\}_{c>0}$, the 1-semi-additive functional equation is

$$g_{c_1c_2}(s) = \frac{g_{c_1}(s)}{c_2} + g_{c_2}(\psi_{c_1}(s)), \quad s\in S,\ c_1,c_2>0. \tag{2.98}$$

The right-hand side of (2.97) is not symmetric in h_1 and h_2. To solve these equations in a special case, we will use the fact that the right-hand side should not change if we permute h_1 and h_2. See, for example, Example 2.16 below.

Definition 2.7. A *2-semi-additive functional* $\{j_h\}_{h\in\mathbb{R}}$ for a measurable nonsingular flow $\{\phi_h\}_{h\in\mathbb{R}}$ is a measurable function $j_h : \mathbb{R}\times S\to\mathbb{R}$ satisfying

$$j_{h_1+h_2}(s) = e^{-\beta_1 h_2} j_{h_1}(s) + \varepsilon_{h_1}(s)\Big\{\frac{d(m\circ\phi_{h_1})}{dm}(s)\Big\}^{\beta_2} j_{h_2}(\phi_{h_1}(s)), \quad s\in S,\ h_1,h_2\in\mathbb{R}, \tag{2.99}$$

where $\beta_1,\beta_2\in\mathbb{R}$ are fixed and $\{\varepsilon_h\}_{h\in\mathbb{R}}$ is a cocycle for the flow $\{\phi_h\}_{h\in\mathbb{R}}$. If (2.99) holds a.e. $m(ds)$ for each fixed $h_1,h_2\in\mathbb{R}$, the functional $\{j_h\}_{h\in\mathbb{R}}$ is called an *almost 2-semi-additive functional*. For a multiplicative flow $\{\psi_c\}_{c>0}$, the 2-semi-additive functional equation is

$$j_{c_1c_2}(s) = c_2^{-\beta_1} j_{c_1}(s) + \varepsilon_{c_1}(s)\Big\{\frac{d(m\circ\psi_{c_1})}{dm}(s)\Big\}^{\beta_2} j_{c_2}(\psi_{c_1}(s)), \quad s\in S,\ c_1,c_2>0. \tag{2.100}$$

The functionals above are named after semi-additive functionals $\{k_h\}_{h\in\mathbb{R}}$ satisfying

$$k_{h_1+h_2}(s) = k_{h_1}(s) + k_{h_2}(\phi_{h_1}(s)),$$

which were studied by Kubo [28]. The Jacobian

$$\frac{d(m\circ\phi_h)}{dm}(s)$$

in (2.99) is defined, in general, only a.e. $m(ds)$. Moreover, it is always an almost cocycle taking values in $\mathbb{R}$ (see Remark 2.12) since

$$\frac{d(m\circ\phi_{h_1+h_2})}{dm}(s) = \frac{d(m\circ\phi_{h_1}\circ\phi_{h_2})}{dm}(s) = \frac{d(m\circ\phi_{h_1}\circ\phi_{h_2})}{d(m\circ\phi_{h_2})}(s)\frac{d(m\circ\phi_{h_2})}{dm}(s)$$

$$= \frac{d(m\circ\phi_{h_1})}{dm}(\phi_{h_2}(s))\frac{d(m\circ\phi_{h_2})}{dm}(s) \quad \text{a.e. } m(ds).$$

In the relation (2.99), this Jacobian is assumed to be chosen so that it is jointly measurable in h and s, and is a cocycle taking values in $\mathbb{R}$ (not just an almost cocycle). This will be possible in the context of stable processes considered below (see Step 2 in the proof of Theorem 3.1).

In the next three examples, we characterize 1- and 2-semi-additive functionals associated with dissipative, fixed, and cyclic flows.

Example 2.16. If $\{\phi_h\}_{h\in\mathbb{R}}$ is a fixed flow (2.71), we have $\phi_h(s) = s$ and the cocycle equation (2.97) defining 1-semi-additive functionals becomes

$$g_{h_1+h_2}(s) = e^{-h_2}g_{h_1}(s) + g_{h_2}(s) = e^{-h_1}g_{h_2}(s) + g_{h_1}(s), \quad s\in S,\ h_1,h_2\in\mathbb{R}.$$

This implies that

$$(e^{-h_2}-1)g_{h_1}(s) = (e^{-h_1}-1)g_{h_2}(s).$$

We can solve this equation for g_h. Set $h_1 = h$, $h_2 = 1$, yielding

$$g_h(s) = (e^{-h}-1)g(s), \tag{2.101}$$

where $g(s) = (e^{-1}-1)^{-1}g_1(s)$.

Turning to 2-semi-additive functionals satisfying (2.99), we can take

$$\frac{d(m\circ\phi_h)}{dm}(s) = \frac{dm}{dm}(s) \equiv 1$$

as a cocycle for the fixed flow $\{\phi_h\}_{h\in\mathbb{R}}$. By using Example 2.13, a cocycle for the fixed flow is $\varepsilon_h(s) = 1$. The equation (2.99) becomes

$$j_{h_1+h_2}(s) = e^{-\beta_1 h_2}j_{h_1}(s) + j_{h_2}(s), \quad s\in S,\ h_1,h_2\in\mathbb{R}.$$

By using the same argument as above when $\beta_1 \neq 0$ and Theorem 1.1.8 in Bingham et al. [5] when $\beta_1 = 0$, we have

$$j_h(s) = \begin{cases} (e^{-\beta_1 h}-1)j(s), & \text{if } \beta_1 \neq 0,\\ h\,j(s), & \text{if } \beta_1 = 0,\end{cases} \tag{2.102}$$

for some function j.

Example 2.17. By Theorem 2.3, any dissipative flow is null-isomorphic to a dissipative flow $\{\widetilde{\phi}_h\}_{h\in\mathbb{R}}$ given in (2.81). If $\{\widetilde{g}_h\}_{h\in\mathbb{R}}$ is a 1-semi-additive functional associated with the flow $\{\widetilde{\phi}_h\}_{h\in\mathbb{R}}$, the equation (2.97) becomes

$$\widetilde{g}_{h_1+h_2}(y,u) = e^{-h_2}\widetilde{g}_{h_1}(y,u) + \widetilde{g}_{h_2}(y,u+h_1), \quad (y,u) \in Y \times \mathbb{R},\ h_1,h_2 \in \mathbb{R}.$$

To solve this equation for $\widetilde{g}_h$, set $u = 0$, $h_2 = h$ and $h_1 = v$. This leads to

$$\widetilde{g}_{v+h}(y,0) = e^{-h}\widetilde{g}_v(y,0) + g_h(y,v).$$

Setting $\widetilde{g}(y,v) = \widetilde{g}_v(y,0)$, we obtain that

$$\widetilde{g}_h(y,v) = \widetilde{g}(y,v+h) - e^{-h}\widetilde{g}(y,v) \tag{2.103}$$

for a function $\widetilde{g} : Y \times \mathbb{R} \to \mathbb{R}$. For a multiplicative flow, a 1-semi-additive functional satisfies

$$\widetilde{g}_c(y,s) = \widetilde{g}(y,s+\ln c) - c^{-1}\widetilde{g}(y,s), \tag{2.104}$$

where $\widetilde{g}$ is some measurable function.

Turning to 2-semi-additive functionals, let $\mathbb{L}$ denote the Lebesgue measure on $\mathbb{R}$. We can take

$$\frac{d((\nu \times \mathbb{L}) \circ \widetilde{\phi}_h)}{d(\nu \times \mathbb{L})}(y,u) = \frac{d\nu(y)}{d\nu(y)}\frac{d(u+h)}{du} \equiv 1$$

as a cocycle for the dissipative flow $\{\widetilde{\phi}_h\}_{h\in\mathbb{R}}$ in (2.81). By Example 2.14, the cocycle $\widetilde{\varepsilon}_h$ is given by

$$\widetilde{\varepsilon}_h(y,u) = \frac{\widetilde{\varepsilon}(y,u+h)}{\widetilde{\varepsilon}(y,u)}$$

with some function $\widetilde{\varepsilon}$ taking values in $\{-1,1\}$. Hence, a 2-semi-additive functional $\{\widetilde{j}_h\}_{h\in\mathbb{R}}$ for the flow $\{\widetilde{\phi}_h\}_{h\in\mathbb{R}}$ satisfies

$$\widetilde{j}_{h_1+h_2}(y,u) = e^{-\beta_1 h_2}\widetilde{j}_{h_1}(y,u) + \frac{\widetilde{\varepsilon}(y,u+h)}{\widetilde{\varepsilon}(y,u)}\widetilde{j}_{h_2}(y,u+h)$$

for all $(y,u) \in Y \times \mathbb{R}$ and $h_1,h_2 \in \mathbb{R}$.

To solve this equation for $\widetilde{j}_h$, set

$$j_h(y,u) = \widetilde{\varepsilon}(y,u)\widetilde{j}_h(y,u)$$

so that

$$j_{h_1+h_2}(y,u) = e^{-\beta_1 h_2} j_{h_1}(y,u) + j_{h_2}(y,u+h_1).$$

Substituting $u = 0$ into this relation, setting $h_1 = v$, $h_2 = h$, and using

$$j_{v+h}(y,0) = j(y,v+h),$$

we obtain that

$$j_h(y,v) = j(y,v+h) - e^{-\beta_1 h} j(y,v).$$

Hence,

$$\widetilde{j}_h(y,u) = (\widetilde{\varepsilon}(y,u))^{-1} j_h(y,u) = \frac{j(y,u+h)}{\widetilde{\varepsilon}(y,u)} - e^{-\beta_1 h}\frac{j(y,u)}{\widetilde{\varepsilon}(y,u)}$$

$$= \frac{\widetilde{\varepsilon}(y,u+h)}{\widetilde{\varepsilon}(y,u)}\widetilde{j}(y,u+h) - e^{-\beta_1 h}\widetilde{j}(y,u), \tag{2.105}$$

where $\widetilde{j}(y,u) = j(y,u)/\widetilde{\varepsilon}(y,u)$ is some function.

For a multiplicative flow, a 2-semi-additive functional satisfies

$$\widetilde{j}_c(y,s) = \frac{\widetilde{b}(y,s+\ln c)}{\widetilde{b}(y,s)}\widetilde{j}(y,s+\ln c) - c^{-\beta_1}\widetilde{j}(y,u), \tag{2.106}$$

where $\widetilde{j}$ is some measurable function.

Example 2.18. By Proposition 2.7, any cyclic flow is null-isomorphic to a cyclic flow $\{\widetilde{\phi}_h\}_{h\in\mathbb{R}}$ given in (2.87). If $\{\widetilde{g}_h\}_{h\in\mathbb{R}}$ is a 1-semi-additive functional associated with the flow $\{\widetilde{\phi}_h\}_{h\in\mathbb{R}}$, one can show that

$$\widetilde{g}_h(z,u) = \widetilde{g}(z,\{u+h\}_{q(z)}) - e^{-h}\widetilde{g}(z,u), \tag{2.107}$$

for a function $\widetilde{g} : Z\times[0,q(\cdot))\to\mathbb{R}$ (Pipiras and Taqqu [41], Proposition 5.1). If $\{\widetilde{j}_h\}_{h\in\mathbb{R}}$ is a 2-semi-additive functional associated with the flow $\{\widetilde{\phi}_h\}_{h\in\mathbb{R}}$, one can show that

$$\widetilde{j}_h(z,v) = \widetilde{\varepsilon}_0(z)^{[v+h]_{q(z)}}\frac{\widetilde{\varepsilon}(z,\{u+h\}_{q(z)})}{\widetilde{\varepsilon}(z,v)}\widetilde{j}(z,\{u+h\}_{q(z)}) - e^{-\beta_1 h}\widetilde{j}(z,v)$$

$$+\frac{\widetilde{j}_0(z)}{\widetilde{\varepsilon}(z,v)}[v+h]_{q(z)}1_{\{\widetilde{\varepsilon}_0(z)=1\}}1_{\{\beta_1=0\}}, \tag{2.108}$$

for some functions $\widetilde{j}_1 : Z\to\mathbb{R}$ and $\widetilde{j} : Z\times[0,q(\cdot))\to\mathbb{R}$, where $\widetilde{\varepsilon}$ and $\widetilde{\varepsilon}_0$ are the functions appearing in Example 2.15 (Pipiras and Taqqu [41], Proposition 5.2).

The next result is analogous to Theorem 2.4.

Proposition 2.8. *An almost 1-semi-additive (2-semi-additive, resp.) functional has a modification which is a 1-semi-additive (2-semi-additive, resp.) functional.*

The proposition is proved in Pipiras and Taqqu [41] (see Theorem 2.1, and Examples 3.1 and 3.2) following the approach of Kubo [28] based on special flows.

Chapter 3
Mixed Moving Averages and Self-Similarity

3.1 Self-Similar Mixed Moving Averages

We will focus on a large subclass of $S\alpha S$, self-similar processes with stationary increments, known as self-similar mixed moving averages. Mixed moving averages are defined next.

Definition 3.1. (Stationary increments) *Mixed moving average*[1] is a $S\alpha S$ process $X_\alpha(t), t \in \mathbb{R}$, having an integral representation

$$\int_X \int_{\mathbb{R}} (G(x,t+u) - G(x,u))M(dx,du), \tag{3.1}$$

where $M(dx,du)$ is a $S\alpha S$ random measure on $X \times \mathbb{R}$ with control measure $\mu(dx)du$, and

$$G_t(x,u) = G(x,t+u) - G(x,u) \in L^\alpha(X \times \mathbb{R}, \mu(dx)du). \tag{3.2}$$

A mixed moving average has always stationary increments by construction. In the Gaussian case $\alpha = 2$, under mild assumptions, all Gaussian stochastic processes with stationary increments (and 0 at $t = 0$) can be written as in (3.1) without X.[2] The stable case is different. The mixing space X matters and, in fact, not all stable processes with stationary increments are mixed moving averages.

Consider now those mixed moving averages that are also *self-similar*. The self-similarity relation $c^{-H}X_\alpha(ct) \stackrel{d}{=} X_\alpha(t), t \in \mathbb{R}$, implies

[1]The process is called "mixed" because of the presence of the X dimension which gives rise to a "mixture."

[2]Indeed, under mild assumptions (e.g., Doob [11], Section 11, Chapter XI), such a process has a spectral domain representation $\int_{\mathbb{R}} \widehat{f}_t(x)\widehat{B}(dx)$, where $\widehat{f}_t(x) = (e^{itx} - 1)f(x)$ and $\widehat{B}(dx)$ is a Gaussian Hermitian measure with the control measure dx. Then, under additional mild assumptions, the postulated representation $\int_{\mathbb{R}}(G(t+u) - G(u))B(du)$ follows by setting G to be the inverse Fourier transform of the function f.

V. Pipiras, M.S. Taqqu, *Stable Non-Gaussian Self-Similar Processes with Stationary Increments*, SpringerBriefs in Probability and Mathematical Statistics,
DOI 10.1007/978-3-319-62331-3_3

$$\int_X \int_{\mathbb{R}} (G(x,t+u) - G(x,u))M(dx,du) \stackrel{d}{=} \int_X \int_{\mathbb{R}} c^{-H}(G(x,ct+u) - G(x,u))M(dx,du)$$

$$= \int_X \int_{\mathbb{R}} c^{-H}(G(x,c(t+u)) - G(x,cu))M(dx,dcu)$$

$$\stackrel{d}{=} \int_X \int_{\mathbb{R}} c^{-H+1/\alpha}(G(x,c(t+u)) - G(x,cu))M(dx,du) \tag{3.3}$$

or, equivalently, for any $\theta_1,\dots,\theta_n \in \mathbb{R}$, $t_1,\dots,t_n \in \mathbb{R}$, $n \geq 1$,

$$\int_X \int_{\mathbb{R}} \left| \sum_{k=1}^{n} \theta_k (G(x,t_k+u) - G(x,u)) \right|^{\alpha} \mu(dx)du$$

$$= \int_X \int_{\mathbb{R}} \left| \sum_{k=1}^{n} \theta_k c^{-H\alpha+1} (G(x,c(t_k+u)) - G(x,cu)) \right|^{\alpha} \mu(dx)du. \tag{3.4}$$

As the next examples show, self-similarity obviously imposes conditions on the types of functions G that can be chosen.

Example 3.1. LFSM in Example 2.5 is a self-similar mixed moving averages with $X = \{1\}$, $\mu(dx) = \delta_{\{1\}}(dx)$ and

$$G(1,u) = au_+^{H-1/\alpha} + bu_-^{H-1/\alpha}.$$

Example 3.2. Let $\alpha \in (1,2)$ and $H \in (1/\alpha, 1)$. Consider the $S\alpha S$ process given by the integral representation

$$\int_0^{\infty} \int_{\mathbb{R}} \Big((t-u)_+ \wedge x - (-u)_+ \wedge x \Big) x^{H-\frac{2}{\alpha}-1} M(dx,du), \tag{3.5}$$

where $M(dx,du)$ is a $S\alpha S$ random measure on $\mathbb{R} \times (0,\infty)$ with the control measure $dudx$. The process can be checked to be well defined, H-self-similar, and have stationary increments. It is called the *Telecom process*, since it arises as the limit of processes modeling Internet traffic (e.g., Pipiras and Taqqu [42]). Note that the Telecom process is a self-similar mixed moving average with $X = (0,\infty)$, $\mu(dx) = dx$ and

$$G(x,u) = (u_+ \wedge x) x^{H-\frac{2}{\alpha}-1}.$$

Since $u_+ \wedge x = ((u-x) \wedge 0 + x)_+$, by making the change of variables $u - x$ to u in the mixed moving average representation, the Telecom process is also a self-similar mixed moving averages with $X = (0,\infty)$, $\mu(dx) = dx$ and

$$G(x,u) = (u \wedge 0 + x)_+ x^{H-\frac{2}{\alpha}-1}.$$

The last representation of the Telecom process will be used in the sequel.

3.1.1 Minimality

We include here a number of results on *minimal representations* and consequences of the rigidity properties for mixed moving averages. We do so for completeness and its own interest rather than for later use, since the theory developed in Section 3.2 below will generally apply to mixed moving average representations which are not necessarily minimal.

The next example is an analogue of Example 2.2 in the case of stationary increments.

Example 3.3. Let X_α be a (nondegenerate) $S\alpha S$ process with stationary increments, having the representation

$$\{X_\alpha(t)\}_{t\in\mathbb{R}} \stackrel{d}{=} \left\{\int_{\mathbb{R}} (G(t+u)-G(u))M(du)\right\}_{t\in\mathbb{R}}, \tag{3.6}$$

where M has the Lebesgue control measure du. The representations of the LFSM (Example 3.1) and log-fractional process (Example 3.7 below) are special cases of (3.6). The process X_α is a mixed moving average with $X = \{1\}$, $\mu(dx) = \delta_{\{1\}}(dx)$. Denote

$$G_t(u) = G(t+u) - G(u), \quad u,t \in \mathbb{R}.$$

We will show that the integral representation $\{G_t\}_{t\in\mathbb{R}}$ of X_α is minimal.

We shall not prove that the integral representation $\{G_t\}_{t\in\mathbb{R}}$ has full support as in (1.25) (see Example 4.1 in Pipiras and Taqqu [35]) but focus rather on the condition (2.4)–(2.5) involving minimality. We need to show that if for some measurable nonsingular map $\phi : \mathbb{R} \to \mathbb{R}$, a measurable function $k : \mathbb{R} \to \mathbb{R} \setminus \{0\}$ and all $t \in \mathbb{R}$,

$$G_t(u) = G(t+u) - G(u) = k(u)(G(t+\phi(u)) - G(\phi(u))) = k(u)G_t(\phi(u)) \quad \text{a.e. } du, \tag{3.7}$$

then $\phi(u) = u$ a.e. du. By replacing t by $t+v$ in (3.7), we get, for all $t, v \in \mathbb{R}$,

$$G(t+v+u) - G(u) = k(u)(G(t+v+\phi(u)) - G(\phi(u))) \quad \text{a.e. } du \tag{3.8}$$

and, by subtracting (3.7) from (3.8),

$$G(t+v+u) - G(t+u) = k(u)(G(t+v+\phi(u)) - G(t+\phi(u))) \quad \text{a.e. } du. \tag{3.9}$$

By Lemma 1.1, the relation (3.9) holds also a.e. $dtdvdu$. By making the change of variables $t+u = z$, we then get

$$G(v+z) - G(z) = k(u)(G(v+z+\phi(u)-u) - G(z+\phi(u)-u)) \quad \text{a.e. } dzdvdu. \tag{3.10}$$

Suppose by contradiction that $\phi \neq Id$ a.e. Then, there is u_0 such that $\delta_0 = u_0 - \phi(u_0) \neq 0$ and

$$G(v+z) - G(z) = k(u_0)(G(v+z+\phi(u_0)-u_0) - G(z+\phi(u_0)-u_0))$$

$$= k(u_0)(G(v+z+\delta_0) - G(z+\delta_0)) \quad \text{a.e. } dvdz. \tag{3.11}$$

But then, arguing as in (2.14), a.e. dv,

$$\int_{\mathbb{R}} |G(v+z) - G(z)|^\alpha dz = \sum_{n=-\infty}^{\infty} |k(u_0)|^\alpha \int_{(n-1)\delta_0}^{n\delta_0} |G(v+z-\delta_0) - G(z-\delta_0)|^\alpha dz$$

$$= \sum_{n=-\infty}^{\infty} |k(u_0)|^\alpha \int_{(n-2)\delta_0}^{(n-1)\delta_0} |G(v+z) - G(z)|^\alpha dz$$

$$= \sum_{n=-\infty}^{\infty} |k(u_0)|^{2\alpha} \int_{(n-2)\delta_0}^{(n-1)\delta_0} |G(v+z-\delta_0) - G(z-\delta_0)|^\alpha dz$$

$$= \ldots = \int_0^{\delta_0} |G(v+z) - G(z)|^\alpha dz \sum_{n=-\infty}^{\infty} |k(u_0)|^{\alpha n} = \infty,$$

unless $G(v+z) - G(z) = 0$ a.e. $dvdz$. The latter condition, however, implies that after the change of variables $v+z = w$,

$$G(w) = G(z) \quad \text{a.e. } dwdz.$$

Hence, by fixing w_0 for which this relation holds a.e. dz, we get that $G(z) = \text{const}$ a.e. dz, which contradicts the assumption that the process is nondegenerate. Hence, $\phi = Id$ a.e. and therefore by Definition 2.1, the representation is minimal. We thus have proved the following result.

Proposition 3.1. *The representations of LFSM (Example 3.1) and log-fractional stable motion (Example 3.7 below) are minimal.*

In the next example, we also show the minimality of the representation (3.5) of the Telecom process in Example 3.2.

Example 3.4. Consider the representation (3.5) of the Telecom process having the kernel function

$$G_t(x,u) = \Big((t-u)_+ \wedge x - (-u)_+ \wedge x\Big) x^{H-\frac{2}{\alpha}-1} \tag{3.12}$$

on $(0,\infty)\times\mathbb{R}$. We will show that the representation $\{G_t\}_{t\in\mathbb{R}}$ is minimal. By (2.4)–(2.5), it is enough to show that for each $t \in \mathbb{R}$,

$$G_t(x,u) = h(x,u) G_t(\phi(x,u)) \quad \text{a.e. } dxdu \tag{3.13}$$

for nonsingular $\phi(x,u) = (\phi_1(x,u), \phi_2(x,u)) : (0,\infty)\times\mathbb{R} \to (0,\infty)\times\mathbb{R}$ and $h : (0,\infty)\times\mathbb{R} \to \mathbb{R}\setminus\{0\}$, entails $\phi(x,u) = (x,u)$ a.e. $dxdu$, that is,

$$\phi_1(x,u) = x, \ \phi_2(x,u) = u \quad \text{a.e. } dxdu. \tag{3.14}$$

In view of (3.12), the relation (3.13) can be expressed as: for $t \in \mathbb{R}$,

$$\Big((t-u)_+\wedge x-(-u)_+\wedge x\Big)x^{H-\frac{2}{\alpha}-1}$$

$$=h(x,u)\Big((t-\phi_2(x,u))_+\wedge\phi_1(x,u)-(-\phi_2(x,u))_+\wedge\phi_1(x,u)\Big)\phi_1(x,u)^{H-\frac{2}{\alpha}-1} \tag{3.15}$$

a.e. $dxdu$. By Lemma 1.1, (i), the relation (3.15) holds a.e. $dtdxdu$. By making the change of variables $v=t-u$, $u=u$, and $x=x$, we also have

$$\Big(v_+\wedge x-(-u)_+\wedge x\Big)x^{H-\frac{2}{\alpha}-1}$$

$$=h(x,u)\Big((v+u-\phi_2(x,u))_+\wedge\phi_1(x,u)-(-\phi_2(x,u))_+\wedge\phi_1(x,u)\Big)\phi_1(x,u)^{H-\frac{2}{\alpha}-1} \tag{3.16}$$

a.e. $dvdxdu$. By considering the two sides of (3.16) as piecewise linear functions of v, we deduce that (3.16) holds only if $u-\phi_2(x,u)=0$, $x^{H-2/\alpha-1}=h(x,u)\phi_1(x,u)^{H-2/\alpha-1}$, and $\phi_1(x,u)=x$ a.e. $dxdu$. In particular, the relations in (3.14) hold.

In view of the preceding example, it should perhaps not be surprising that a mixed moving average always has a minimal mixed-moving average representation, possibly with a different underlying mixing space X.

Proposition 3.2. *Let $\alpha\in(0,2)$, $H>0$ and X_α be a SαS mixed moving average process with stationary increments given by (3.1) and (3.2). Then, there exists a standard Lebesgue space $(\widetilde{X},\mathscr{B}(\widetilde{X}),\widetilde{\mu})$ and measurable function*

$$\widetilde{G}:\widetilde{X}\times\mathbb{R}\to\mathbb{R}$$

such that

$$\begin{aligned}\{X_\alpha(t)\}_{t\in\mathbb{R}} &\stackrel{d}{=}\left\{\int_{\widetilde{X}}\int_{\mathbb{R}}(\widetilde{G}(\widetilde{x},t+u)-\widetilde{G}(\widetilde{x},u))\widetilde{M}(d\widetilde{x},du)\right\}_{t\in\mathbb{R}}\\ &=\left\{\int_{\widetilde{X}}\int_{\mathbb{R}}\widetilde{G}_t(\widetilde{x},u)\widetilde{M}(d\widetilde{x},du)\right\}_{t\in\mathbb{R}},\end{aligned}$$

where $\widetilde{M}$ has the control measure $\widetilde{\mu}(d\widetilde{x})du$, and that the integral representation $\{\widetilde{G}_t\}_{t\in\mathbb{R}}$ is minimal for X_α.

We shall not prove Proposition 3.2. When $\alpha\in(1,2)$, it is proved in Pipiras and Taqqu [35], Theorem 4.2, by transforming mixed moving averages to stationary mixed moving averages and using the results of Rosiński [46]. A different approach, based on (non)minimal sets and valid for all $\alpha\in(0,2)$, is taken by Pipiras [34], Theorem 5.1. In particular, the latter result shows that to construct a minimal representation, one can take a subset of the mixing space X.

3.1.2 Rigidity

The next two results express rigidity properties of the representations of mixed moving averages and will be used below.

Proposition 3.3. *Let $\alpha \in (0,2)$ and consider a $S\alpha S$ process $\{X_\alpha(t)\}_{t\in\mathbb{R}}$, given by (3.1) and (3.2), with*

$$\operatorname{supp}\{G_t, t \in \mathbb{R}\} = X \times \mathbb{R} \quad a.e.\ \mu(dx)du.$$

Suppose that $\{X_\alpha(t)\}_{t\in\mathbb{R}}$ has another integral representation

$$\begin{aligned}\{X_\alpha(t)\}_{t\in\mathbb{R}} &\stackrel{d}{=} \left\{\int_{\widetilde{X}}\int_{\mathbb{R}} \widetilde{G}_t(\widetilde{x},u)\widetilde{M}(d\widetilde{x},du)\right\}_{t\in\mathbb{R}} \\ &= \left\{\int_{\widetilde{X}}\int_{\mathbb{R}} (\widetilde{G}(\widetilde{x},t+u) - \widetilde{G}(\widetilde{x},u))\widetilde{M}(d\widetilde{x},du)\right\}_{t\in\mathbb{R}},\end{aligned}$$

where $(\widetilde{X}, \mathscr{B}(\widetilde{X}), \widetilde{\mu})$ is also a standard Lebesgue space,

$$\{\widetilde{G}_t\}_{t\in\mathbb{R}} \subset L^\alpha(\widetilde{X}\times\mathbb{R}, \widetilde{\mu}(d\widetilde{x})du),$$

a $S\alpha S$ random measure $\widetilde{M}$ has the control measure $\mu(d\widetilde{x})du$. Then, there exist measurable functions

$$\Phi_1 : X \to \widetilde{X},\ h : X \to \mathbb{R}\setminus\{0\} \quad and \quad \Phi_2, \Phi_3 : X \to \mathbb{R}$$

such that

$$G(x,u) = h(x)\widetilde{G}\Big(\Phi_1(x), u + \Phi_2(x)\Big) + \Phi_3(x) \tag{3.17}$$

a.e. $\mu(dx)du$.

PROOF: By applying Proposition 2.5, we obtain that there exist functions $\widetilde{\Phi}_1 : X \times \mathbb{R} \to \widetilde{X}$, $\widetilde{h} : X\times\mathbb{R} \to \mathbb{R}\setminus\{0\}$ and $\widetilde{\Phi}_2 : X\times\mathbb{R} \to \mathbb{R}$ such that

$$G(x,t+u) - G(x,u) = \widetilde{h}(x,u)\left(\widetilde{G}(\widetilde{\Phi}_1(x,u), t + \widetilde{\Phi}_2(x,u)) - \widetilde{G}(\widetilde{\Phi}_1(x,u), \widetilde{\Phi}_2(x,u))\right) \tag{3.18}$$

a.e. $\mu(dx)dtdu$. Now, by making the change of variables $v = t+u$ in (3.18), we get that

$$G(x,v) = \widetilde{h}(x,u)\widetilde{G}(\widetilde{\Phi}_1(x,u), v - u + \widetilde{\Phi}_2(x,u)) + \widetilde{\Phi}_3(x,u) \tag{3.19}$$

a.e. $\mu(x)dvdu$, for some measurable function $\widetilde{\Phi}_3$. Now, fix $u = u_0$, for which (3.19) holds a.e. $\mu(dx)dv$ (and for which the functions involved are measurable) to obtain the result of the proposition after the proper change of notation. □

The next result states that the functions in (3.17) are unique and that there is a change of measure formula if the representation $\widetilde{G}_t$ is minimal.

Proposition 3.4. *Assume in Proposition 3.3 that the integral representation*

$$\{\widetilde{G}_t\}_{t\in\mathbb{R}} \subset L^\alpha(\widetilde{X}\times\mathbb{R}, \widetilde{\mu}(d\widetilde{x})du)$$

is minimal and $\operatorname{supp}\{G_t, t\in\mathbb{R}\} = X\times\mathbb{R}$. *Then, there are unique (modulo* μ*) measurable functions*

$$\Phi_1 : X\to\widetilde{X},\ h : X\to\mathbb{R}\setminus\{0\} \quad \text{and} \quad \Phi_2, \Phi_3 : X\to\mathbb{R}$$

such that (3.17) holds a.e. $\mu(dx)du$*, with*

$$\widetilde{\mu} = \mu_h\circ\Phi_1^{-1} \tag{3.20}$$

on $\mathscr{B}(\widetilde{X})$*, where*

$$\mu_h(dx) = |h(x)|^\alpha\mu(dx).$$

PROOF: By Proposition 2.3, there exist unique modulo $\mu(dx)du$ functions $\widetilde{\Phi}_1 : X\times\mathbb{R}\to\widetilde{X}$, $\widetilde{h} : X\times\mathbb{R}\to\mathbb{R}\setminus\{0\}$ and $\widetilde{\Phi}_2 : X\times\mathbb{R}\to\mathbb{R}$ such that the relation (3.18) holds and

$$\widetilde{\mu}(d\widetilde{x})du = (\mu_{\widetilde{h}}\circ\widetilde{\Phi}^{-1})(d\widetilde{x}, du),$$

where $\widetilde{\Phi} = (\widetilde{\Phi}_1, \widetilde{\Phi}_2)$ and

$$\mu_{\widetilde{h}}(dx, du) = |\widetilde{h}(x,u)|^\alpha\mu(dx)du. \tag{3.21}$$

Now, set $s\in\mathbb{R}$. By replacing t by $t+s$ in the relation (3.18) and subtracting the same relation (3.18) with t replaced by s, we obtain that

$$G(x,t+s+u) - G(x,s+u)$$
$$= \widetilde{h}(x,u)\left(\widetilde{G}(\widetilde{\Phi}_1(x,u), t+s+\widetilde{\Phi}_2(x,u)) - \widetilde{G}(\widetilde{\Phi}_1(x,u), s+\widetilde{\Phi}_2(x,u))\right)$$

a.e. $\mu(dx)du$. By Lemma 1.1, the last relation holds a.e. $\mu(dx)duds$ and, by making the change of variables $s+u=z$, we get that

$$G(x,t+z) - G(x,z)$$
$$= \widetilde{h}(x,u)\left(\widetilde{G}(\widetilde{\Phi}_1(x,u), t+z+\widetilde{\Phi}_2(x,u)-u) - \widetilde{G}(\widetilde{\Phi}_1(x,u), z+\widetilde{\Phi}_2(x,u)-u)\right)$$

a.e. $\mu(dx)dudz$. Finally, fix $u=u_0$ for which this relation holds a.e. $\mu(dx)dz$ so that

$$G(x,t+z) - G(x,z) = h(x)\left(\widetilde{G}(\Phi_1(x), t+z+\Phi_2(x)) - \widetilde{G}(\Phi_1(x), z+\Phi_2(x))\right) \tag{3.22}$$

a.e. $\mu(dx)dz$, where

$$\Phi_1(x) = \widetilde{\Phi}_1(x,u_0),\ \Phi_2(x) = \widetilde{\Phi}_2(x,u_0) - u_0 \quad \text{and} \quad h(x) = \widetilde{h}(x,u_0).$$

Now replace z by u in (3.22) so that it can be easily compared with (3.18). The uniqueness of the functions in (3.18) yields

$$\widetilde{\Phi}_1(x,u) = \Phi_1(x), \quad \widetilde{\Phi}_2(x,u) = u + \Phi_2(x), \quad \widetilde{h}(x,u) = h(x)$$

a.e. $\mu(dx)du$. Since, for $A \in \mathscr{B}(X)$ and $B \in \mathscr{B}(\mathbb{R})$, by using (3.21),

$$\int 1_{A\times B}(\widetilde{\Phi}(x,u))\mu_{\widetilde{h}}(dx,du) = \int 1_{A\times B}(\widetilde{\Phi}_1(x,u),\widetilde{\Phi}_2(\widetilde{x},u))|\widetilde{h}(x,u)|^{\alpha}\mu(dx)du$$

$$= \int 1_{A\times B}(\Phi_1(x),\Phi_2(x)+u)|h(x)|^{\alpha}\mu(dx)du = \int 1_A(\Phi_1(x))1_B(u)|h(x)|^{\alpha}\mu(dx)du,$$

it follows that

$$(\mu_{\widetilde{h}} \circ \widetilde{\Phi}^{-1})(d\widetilde{x},du) = (\mu_h \circ \Phi_1^{-1})(d\widetilde{x})du,$$

where $\mu_h(dx) = |h(x)|^{\alpha}\mu(dx)$. The relation (3.17) follows from (3.18) as in the proof of Proposition 3.3. □

Proposition 3.5. *Assume in Proposition 3.3 that both integral representations* $\{G_t\}_{t\in\mathbb{R}}$ *and* $\{\widetilde{G}_t\}_{t\in\mathbb{R}}$ *are minimal. Then, there are unique (modulo* μ*) null-isomorphism*

$$\Phi_1 : X \to \widetilde{X},\ \varepsilon : X \to \{-1,1\} \quad and \quad \Phi_2,\Phi_3 : X \to \mathbb{R}$$

such that (3.17) holds a.e. $\mu(dx)du$*, with*

$$h(x) = \varepsilon(x)\left\{\frac{d(\mu\circ\Phi_1)}{d\mu}(x)\right\}^{1/\alpha},$$

that is,

$$G(x,u) = \varepsilon(x)\left\{\frac{d(\mu\circ\Phi_1)}{d\mu}(x)\right\}^{1/\alpha}\widetilde{G}\Big(\Phi_1(x),u+\Phi_2(x)\Big)+\Phi_3(x). \tag{3.23}$$

PROOF: The proof follows that of Proposition 3.5 above but using Proposition 2.4, instead of Proposition 2.3. Relation (3.23) follows from (3.17). □

3.2 Structure of Self-Similar Mixed Moving Averages

In Section 3.2.1, we relate self-similar mixed moving averages to flows. Based on the Hopf decomposition of the underlying flows, we decompose self-similar mixed moving averages into two classes in Section 3.2.2. The first class, consisting of processes related to dissipative flows, is examined closer in Sections 3.2.3 and 3.2.4. The second class, consisting of processes related to conservative flows, is considered in Sections 3.2.5, 3.2.6, and 3.2.8, where a finer decomposition is obtained. The results are summarized in Section 3.2.7.

3.2.1 Connection to Multiplicative Flows

We first provide a fundamental definition relating self-similar mixed moving averages to flows. The definition is motivated by Theorem 3.1 below, which establishes this relationship for minimal representations. Throughout this section, we let $\alpha \in (0,2)$, $H > 0$ and also set

$$\kappa = H - \frac{1}{\alpha}. \tag{3.24}$$

We also suppose that $(X, \mathscr{B}(X), \mu)$ is a standard Lebesgue space and assume that X_α is H-self-similar mixed moving average with stationary increments. We will now be using *multiplicative flows* instead of additive flows. They are introduced in Section 2.3.2, denoted ψ_c, $c > 0$, and satisfy (2.69).

Definition 3.2. A $S\alpha S$ H-self-similar process with stationary increments, given by a mixed moving average representation (3.1) with a kernel function

$$G_t(x,u) = G(x,t+u) - G(x,u)$$

in (3.2), is *generated by a nonsingular measurable multiplicative flow* $\{\psi_c\}_{c>0}$ on (X,μ) if, for all $c > 0$,

$$c^{-\kappa} G(x,cu) = b_c(x) \left\{ \frac{d(\mu \circ \psi_c)}{d\mu}(x) \right\}^{1/\alpha} G(\psi_c(x), u + g_c(x)) + j_c(x) \tag{3.25}$$

a.e. $\mu(dx)du$, where $\{b_c\}_{c>0}$ is a cocycle for the flow $\{\psi_c\}_{c>0}$ taking values in $\{-1,1\}$, $\{g_c\}_{c>0}$ is a 1-semi-additive functional for the flow $\{\psi_c\}_{c>0}$, $\{j_c\}_{c>0}$ is a 2-semi-additive functional for the flow $\{\psi_c\}_{c>0}$ (with $\beta_1 = \kappa$, $\beta_2 = 1/\alpha$ in (2.100)), and

$$\operatorname{supp}\{G_t, t \in \mathbb{R}\} = X \times \mathbb{R} \quad \text{a.e. } \mu(dx)du. \tag{3.26}$$

Remark 3.1. We do not assume in Definition 3.2 that the representation $\{G_t\}$ is minimal. That is, as far as the definition goes, the representation can be minimal or not. This is convenient since checking for minimality is not necessarily easy. On the other hand, we will also show in Theorem 3.1 below that every minimal representation is related to a flow in the sense of Definition 3.2.

Remark 3.2. The relation (3.25) plays an important role in the sequel and will be often used. It implies that for all $t \in \mathbb{R}$ and $c > 0$,

$$c^{-\kappa}(G(x,c(t+u)) - G(x,cu))$$

$$= b_c(x) \left\{ \frac{d(\mu \circ \psi_c)}{d\mu}(x) \right\}^{1/\alpha} (G(\psi_c(x), t+u+g_c(x)) - G(\psi_c(x), u+g_c(x))) \tag{3.27}$$

a.e. $\mu(dx)du$. In contrast to (3.25), the relation (3.27) does not involve a 2-semi-additive functional $\{j_c\}_{c>0}$.

Remark 3.3. Note also that the relation (3.27) implies the H-self-similarity of the process

$$X_\alpha(t) = \int_X \int_{\mathbb{R}} (G(x,t+u) - G(x,u))M(dx,du).$$

To show that

$$c^{-H} X_\alpha(ct) \stackrel{d}{=} X_\alpha(t),$$

we focus on the exponents of characteristic functions. The self-similarity follows from observing that for any $\theta_1, \dots, \theta_n \in \mathbb{R}$, $t_1, \dots, t_n \in \mathbb{R}$, $n \geq 1$, and $c > 0$,

$$\int_X \int_{\mathbb{R}} \Big| \sum_{k=1}^n \theta_k c^{-H} (G(x, ct_k + u)) - G(x,u)) \Big|^\alpha \mu(dx)du$$

$$= \int_X \int_{\mathbb{R}} \Big| \sum_{k=1}^n \theta_k c^{-H} (G(x, c(t_k + u))) - G(x, cu)) \Big|^\alpha \mu(dx)dcu$$

$$= \int_X \int_{\mathbb{R}} \Big| \sum_{k=1}^n \theta_k c^{-\kappa} (G(x, c(t_k + u)) - G(x, cu)) \Big|^\alpha \mu(dx)du$$

$$= \int_X \int_{\mathbb{R}} \Big| \sum_{k=1}^n \theta_k (G(\psi_c(x), t_k + u + g_c(x)) - G(\psi_c(x), u + g_c(x))) \Big|^\alpha \times$$

$$\times \frac{d(\mu \circ \psi_c)}{d\mu}(x)\, \mu(dx)du$$

$$= \int_X \int_{\mathbb{R}} \Big| \sum_{k=1}^n \theta_k (G(\psi_c(x), t_k + u + g_c(x)) - G(\psi_c(x), u + g_c(x))) \Big|^\alpha (\mu \circ \psi_c)(dx)du$$

$$= \int_X \int_{\mathbb{R}} \Big| \sum_{k=1}^n \theta_k (G(\psi_c(x), t_k + u) - G(\psi_c(x), u) \Big|^\alpha (\mu \circ \psi_c)(dx)du$$

$$= \int_X \int_{\mathbb{R}} \Big| \sum_{k=1}^n \theta_k (G(x, t_k + u) - G(x,u)) \Big|^\alpha \mu(dx)du,$$

where we also used the fact that $\psi_c(X) = X$ (since each ψ_c is one-to-one and onto, with the measurable inverse $\psi_{c^{-1}}$). This shows that the characteristic functions of $(c^{-H}X_\alpha(ct_1), \dots, c^{-H}X_\alpha(ct_n))$ and $(X_\alpha(t_1), \dots, X_\alpha(t_n))$ are identical and hence that $\{c^{-H}X_\alpha(ct)\} \stackrel{d}{=} \{X_\alpha(t)\}$.

Remark 3.4. To get a feeling for the flow, cocycle, and semi-additive functionals appearing in (3.25), let $c_1, c_2 > 0$. On one hand, by (3.25),

$$(c_1c_2)^{-\kappa} G(x, c_1c_2u)$$

$$= b_{c_1c_2}(x) \left\{ \frac{d(\mu \circ \psi_{c_1c_2})}{d\mu}(x) \right\}^{1/\alpha} G(\psi_{c_1c_2}(x), u + g_{c_1c_2}(x)) + j_{c_1c_2}(x). \quad (3.28)$$

On the other hand, by iterating (3.25) twice, one gets

$$(c_1c_2)^{-\kappa}G(x,c_1c_2u) = c_2^{-\kappa}\Big(c_1^{-\kappa}G(x,c_1(c_2u))\Big)$$

$$= c_2^{-\kappa}b_{c_1}(x)\left\{\frac{d(\mu\circ\psi_{c_1})}{d\mu}(x)\right\}^{1/\alpha} G\Big(\psi_{c_1}(x),c_2(u+c_2^{-1}g_{c_1}(x))\Big)+c_2^{-\kappa}j_{c_1}(x)$$

$$= b_{c_1}(x)b_{c_2}(\psi_{c_1}(x))\left\{\frac{d(\mu\circ\psi_{c_1})}{d\mu}(x)\frac{d(\mu\circ\psi_{c_2})}{d\mu}(\psi_{c_1}(x))\right\}^{1/\alpha} \times$$

$$\times G(\psi_{c_2}(\psi_{c_1}(x)),u+c_2^{-1}g_{c_1}(x)+g_{c_2}(\psi_{c_1}(x)))$$

$$+c_2^{-\kappa}j_{c_1}(x)+b_{c_1}(x)\left\{\frac{d(\mu\circ\psi_{c_1})}{d\mu}(x)\right\}^{1/\alpha} j_{c_2}(\psi_{c_1}(x)). \tag{3.29}$$

The flow, cocycle, and semi-additive functionals relations (see (2.69), (2.91), (2.98), and (2.100)) imply that the right-hand sides of (3.28) and (3.29) are identical.

Remark 3.5. Another way of expressing (3.27) is to set

$$\varphi_c(x,u) = \left(\psi_c(x),\frac{u}{c}+g_c(x)\right). \tag{3.30}$$

One can verify that φ_c, $c>0$, is a multiplicative flow on $X\times\mathbb{R}$. Moreover, for all $t>0$,

$$\begin{aligned}
G_t(x,u) &= G(x,t(1+t^{-1}u))-G(x,tt^{-1}u)\\
&= t^{\kappa}\; b_t(x)\left\{\frac{d(\mu\circ\psi_t)}{d\mu}(x)\right\}^{1/\alpha}\times\\
&\qquad\times\left(G(\psi_t(x),1+t^{-1}u+g_t(x))-G(\psi_t(x),t^{-1}u+g_t(x))\right)\\
&= t^{H-\frac{1}{\alpha}}\; b_t(x)\left\{\frac{d(\mu\circ\psi_t)}{d\mu}(x)\right\}^{1/\alpha} G_1(\psi_t(x),t^{-1}u+g_t(x))\\
&= t^{H}\; b_t(x,u)\left\{\frac{d((\mu\times\mathbb{L})\circ\varphi_t)}{d(\mu\times\mathbb{L})}(x)\right\}^{1/\alpha} G_1(\varphi_t(x,u))
\end{aligned} \tag{3.31}$$

a.e. $\mu(dx)du$, where $\mathbb{L}$ denotes the Lebesgue measure and $b_t(x,u)=b_t(x)$. While (3.31) and (3.27) look similar, we shall base our decomposition of the processes *only* on the flow $\psi_t(x)$, that is, on the first component of the flow

$$\varphi_t(x,u) = \left(\psi_t(x),\frac{u}{t}+g_t(x)\right).$$

Example 3.5. Consider LFSM in Example 3.1. We have

$$X=\{1\} \quad\text{and}\quad G_t(1,u)=G(1,t+u)-G(1,u)$$

with
$$G(1,u) = au_+^\kappa + bu_-^\kappa.$$

Since $c^{-\kappa}G(1,cu) = G(1,u)$ and $\text{supp}\{G_t, t \in \mathbb{R}\} = \mathbb{R}$, LFSM is generated by the following flow, related cocycle, and semi-additive functionals, respectively: for all $c > 0$,
$$\psi_c(1) = 1, \quad b_c(1) = 1, \quad g_c(1) = 0, \quad j_c(1) = 0.$$

Example 3.6. Consider the Telecom process in Example 3.2. For $c > 0$, we have that

$$c^{-\kappa}G(x,cu) = c^{-H+\frac{1}{\alpha}}(cu \wedge 0 + x)_+ x^{H-\frac{2}{\alpha}-1}$$
$$= c^{-H+\frac{1}{\alpha}} c c^{H-\frac{2}{\alpha}-1}(u \wedge 0 + c^{-1}x)_+ (c^{-1}x)^{H-\frac{2}{\alpha}-1}$$
$$= c^{-\frac{1}{\alpha}} G(c^{-1}x,u) = \left\{ \frac{d(\mathbb{L} \circ \psi_c)}{d\mathbb{L}}(x) \right\}^{1/\alpha} G(\psi_c(x),u),$$

where $\mathbb{L}$ denotes the Lebesgue measure,

$$\psi_c(x) = c^{-1}x, \quad c > 0, x > 0,$$

is a nonsingular measurable flow on $(0,\infty)$, and where

$$\frac{d(\mathbb{L} \circ \psi_c)}{d\mathbb{L}}(x) = \frac{d(c^{-1}x)}{dx} = c^{-1}.$$

Hence, the condition (3.25) is satisfied with

$$b_c(x) \equiv 1, \quad g_c(x) \equiv 0, \quad j_c(x) \equiv 0.$$

Set
$$G_t(x,u) = G(x,t+u) - G(x,u).$$

Since $\text{supp}\{G_t\} = \{(x,u) : x > 0, -t-x < u < -t\}$ for $t < 0$ and $\text{supp}\{G_t\} = \{(x,u) : x > 0, -t-x < u < 0\}$ for $t > 0$, one has $\text{supp}\{G_t, t \in \mathbb{R}\} = (0,\infty) \times \mathbb{R}$.

Example 3.7. Let $\alpha \in (1,2)$ and $M(du)$ be a $S\alpha S$ random measure on $\mathbb{R}$ with the Lebesgue control measure du. Then,

$$\int_{\mathbb{R}} (\ln|t-u| - \ln|u|) M(du), \quad t \in \mathbb{R}, \tag{3.32}$$

is called a *log-fractional stable motion*. The log-fractional stable motion is a self-similar mixed moving average with

$$X = \{1\},\ \mu = \delta_{\{1\}} \quad \text{and} \quad G(1,u) = \ln|u|.$$

In this case, $\kappa = H - 1/\alpha = 0$ and, for any $c > 0$, $G(1,cu) = G(1,u) + \ln c$. Then, (3.25) is satisfied with

$$\psi_c(1) = 1, \quad b_c(1) = 1, \quad g_c(1) = 0, \quad j_c(1) = \ln c.$$

Definition 3.2 was motivated by the following result. See also Remark 3.1.

Theorem 3.1. *A SαS mixed moving average with a minimal integral representation (3.1) is generated by a flow in the sense of Definition 3.2. Moreover, the flow $\{\psi_c\}_{c>0}$, the cocycle $\{b_c\}_{c>0}$, and the semi-additive functionals $\{g_c\}_{c>0}$ and $\{j_c\}_{c>0}$ are unique modulo μ.*[3]

PROOF: The self-similarity of the process implies that

$$\int_X \int_{\mathbb{R}} c^{-\kappa}(G(x,c(t+u)) - G(x,cu))M(dx,du)$$

$$\stackrel{d}{=} \int_X \int_{\mathbb{R}} (G(x,t+u) - G(x,u))M(dx,du),\ t \in \mathbb{R},$$

in the sense of the finite-dimensional distributions. Since the integral representation $\{G_t\}_{t\in\mathbb{R}}$ is minimal, Proposition 3.5 implies that for every $c > 0$, there are unique modulo $\mu(dx)$ functions

$$\Phi_c^1 : X \to X,\ \Psi_c : X \to \{-1,1\} \quad \text{and} \quad \Phi_c^2, \Phi_c^3 : X \to \mathbb{R}$$

such that Φ_c^1 is (mod 0) null-isomorphism and

$$c^{-\kappa}G(x,cu) = h_c(x)\ G(\Phi_c^1(x), u + \Phi_c^2(x)) + \Phi_c^3(x) \tag{3.33}$$

and

$$h_c(x) = \Psi_c(x)\left\{\frac{d(\mu\circ\Phi_c^1)}{d\mu}(x)\right\}^{1/\alpha} \tag{3.34}$$

a.e. $\mu(dx)du$. We divide the proof into two steps. In *Step 1*, we show that for any $c_1, c_2 > 0$,

$$\Phi_{c_1c_2}^1(x) = \Phi_{c_2}^1(\Phi_{c_1}^1(x)) \quad \text{a.e. } \mu(dx), \tag{3.35}$$

$$\Psi_{c_1c_2}(x) = \Psi_{c_1}(x)\Psi_{c_2}(\Phi_{c_1}^1(x)) \quad \text{a.e. } \mu(dx). \tag{3.36}$$

$$\Phi_{c_1c_2}^2(x) = c_2^{-1}\Phi_{c_1}^2(x) + \Phi_{c_2}^2(\Phi_{c_1}^1(x)) \quad \text{a.e. } \mu(dx), \tag{3.37}$$

$$\Phi_{c_1c_2}^3(x) = c_2^{-\kappa}\Phi_{c_1}^3(x) + \Psi_{c_1}(x)\left\{\frac{d(\mu\circ\Phi_{c_1}^1)}{d\mu}(x)\right\}^{1/\alpha} \Phi_{c_2}^3(\Phi_{c_1}^1(x)) \quad \text{a.e. } \mu(dx). \tag{3.38}$$

[3]Unique modulo μ means, for example, that if $\{\psi_c\}_{c>0}$ and $\{\widetilde{\psi}_c\}_{c>0}$ are two such flows, then for all $c > 0$, $\psi_c = \widetilde{\psi}_c$ μ-a.e.

Modulo the "a.e.", these relations state that Φ^1 is a multiplicative flow, Ψ is a cocycle for the flow Φ^1, Φ^2 is a 1-semi-additive functional associated with Φ^1, and Φ^3 is a 2-semi-additive functional associated with Φ^1. In *Step 2*, we show that the functions Φ_c^1, Ψ_c, Φ_c^2, and Φ_c^3 have versions ψ_c, b_c, g_c, and j_c, respectively, that are measurable in (c,x) and are a flow, cocycle, 1-semi-additive functional, and 2-semi-additive functional, respectively. This will prove the relation (3.25) and conclude the proof. (The uniqueness follows from the stated uniqueness of the maps in (3.33).)

Step 1. Let $c_1, c_2 > 0$. Then, by (3.33),

$$(c_1c_2)^{-\kappa}G(x,c_1c_2u) = h_{c_1c_2}(x)\, G(\Phi^1_{c_1c_2}(x), u + \Phi^2_{c_1c_2}(x)) + \Phi^3_{c_1c_2}(x).$$

On the other hand, by iterating the relation (3.33) twice as in Remark 3.4, we get that[4]

$$\begin{aligned}
&(c_1c_2)^{-\kappa}G(x,c_1c_2u)\\
&= h_{c_1}(x)h_{c_2}(\Phi^1_{c_1}(x))G(\Phi^1_{c_2}(\Phi^1_{c_1}(x)), u + c_2^{-1}\Phi^2_{c_1}(x) + \Phi^2_{c_2}(\Phi^1_{c_1}(x)))\\
&\quad + c_2^{-\kappa}\Phi^3_{c_1}(x) + h_{c_1}(x)\Phi^3_{c_2}(\Phi^1_{c_1}(x)).
\end{aligned}$$

The uniqueness implies that the conditions (3.35) and (3.37) are satisfied. For the condition (3.36), the uniqueness yields

$$h_{c_1c_2}(x) = h_{c_1}(x)h_{c_2}(\Phi^1_{c_1}(x))$$

a.e. $\mu(dx)$ and hence, by using (3.34),

$$\Psi_{c_1c_2}(x)\left\{\frac{d(\mu\circ\Phi^1_{c_1c_2})}{d\mu}(x)\right\}^{1/\alpha}$$

$$= \Psi_{c_1}(x)\left\{\frac{d(\mu\circ\Phi^1_{c_1})}{d\mu}(x)\right\}^{1/\alpha}\Psi_{c_2}(\Phi^1_{c_1}(x))\left\{\frac{d(\mu\circ\Phi^1_{c_2})}{d\mu}(\Phi^1_{c_1}(x))\right\}^{1/\alpha}.$$

The condition (3.36) now follows since

$$\frac{d(\mu\circ\Phi^1_{c_1c_2})}{d\mu} = \frac{d(\mu\circ\Phi^1_{c_2}\circ\Phi^1_{c_1})}{d\mu} = \left(\frac{d(\mu\circ\Phi^1_{c_2})}{d\mu}\circ\Phi^1_{c_1}\right)\frac{d(\mu\circ\Phi^1_{c_1})}{d\mu}.$$

The conditions (3.38) can be proved similarly.

[4]To see this easily, express (3.33) as

$$G(x,u) = h_{c_1}(x)\, c_1^{\kappa}G(\Phi^1_{c_1}(x), c_1^{-1}u + \Phi^2_{c_1}(x)) + c_1^{\kappa}\Phi^3_{c_1}(x)$$

and iterate.

Step 2. We have to deal with joint measurability and a.e. issues. We shall use an argument of Rosiński [46] which applies to flows (modulo the "a.e.") relating integral representations of a process. In the case considered here, the "full" integral representation is

$$G_t(x,u) = G(x,t+u) - G(x,u)$$

defined on the space is $X \times \mathbb{R}$. Our "flow" $\Psi_c^1(x)$, however, is defined only on the space X and relates the function G. We can construct a "flow" related to G_t as in Remark 3.5. To do so, in analogy to (3.30), let

$$F_c(x,u) = \Phi_c\left(x, \frac{u}{c}\right) = \left(\Phi_c^1(x), \frac{u}{c} + \Phi_c^2(x)\right). \tag{3.39}$$

Arguing as for (3.30) (which is left as an exercise), we conclude that, for all $c_1, c_2 > 0$,

$$F_{c_1c_2}(x,u) = F_{c_1}(F_{c_2}(x,u)) \tag{3.40}$$

a.e. $\mu(dx)du$. As in Remark 3.2, we also have

$$c^{-\kappa} G_{ct}(x,u) = h_c(x) G_t(F_c(x,u)) \tag{3.41}$$

a.e. $\mu(dx)du$. We have thus constructed a "flow" F_c satisfying (3.40) which now relates the "full" integral representation G_t in (3.41).

As in the proof of Theorem 3.1 in Rosiński [46], by using a result of Mackey [31], we can conclude from (3.40) and (3.41) that there is a measurable nonsingular multiplicative flow $\{\varphi_c\}_{c>0}$ which is a version of $\{F_c\}_{c>0}$ and that we may also assume without loss of generality that the function $h_c(x)$ in (3.41) is measurable in (c,x). The basic difficulty is to show that we can in fact replace this flow $\{\varphi_c\}_{c>0}$ by the one of the form

$$\left(\psi_c(x), \frac{u}{c} + b_c(x)\right),$$

that is, one of the same form (3.39) as F_c.

We now turn to the task at hand, dealing with the function Φ_c^1 first. Suppose that $\varphi_c(x,u) = (\phi_c^1(x,u), \phi_c^2(x,u))$. Then, for any $c > 0$, $\Phi_c^1(x) = \phi_c^1(x,u)$ a.e. $\mu(dx)du$ and hence

$$\Phi_c^1(x) = \int_0^1 \Phi_c^1(x)du = \int_0^1 \phi_c^1(x,u)du =: \widetilde{\psi}_c(x) \quad \text{a.e. } \mu(dx).$$

Since $\phi_c^1(x,u)$ is measurable in (c,x,u), the function $\widetilde{\psi}_c(x)$ is measurable in (c,x). Moreover, $\{\widetilde{\psi}_c\}_{c>0}$ is a version of $\{\Phi_c^1\}_{c>0}$ and it satisfies $\widetilde{\psi}_{c_1c_2}(x) = \widetilde{\psi}_{c_2}(\widetilde{\psi}_{c_1}(x))$ a.e. $\mu(dx)$ for all $c_1, c_2 > 0$. The point of the preceding argument is that $\widetilde{\psi}_c(x)$ not only satisfies the flow equation a.e. but also that it is jointly measurable in (c,x).

We can now again apply the same arguments as in the proof of Theorem 3.1 in Rosiński [46]. The map (3.7) in that proof, with $\Phi_t(s)$ replaced by $\widetilde{\psi}_{e^t}(x)$, is immediately measurable since $\widetilde{\psi}_{e^t}(x)$ is jointly measurable in (t,x). Then, applying Theorem 1 of Mackey [31] and arguing as at the end of that proof, we can conclude that there is a measurable nonsingular multiplicative flow $\{\psi_c\}_{c>0}$ which is a version of $\{\widetilde{\psi}_c\}_{c>0}$. Observe that $\{\psi_c\}_{c>0}$ is also a version of $\{\Phi_c^1\}_{c>0}$.

Let us now deal with the functions Φ_c^2. Arguing as for the map Φ_c^1 above, we may assume without loss of generality that the map $\Phi_c^2(x) : (0,\infty) \times X \to \mathbb{R}$ is measurable in (c,x). In view of (3.37), it satisfies the relation

$$\Phi_{c_1 c_2}^2(x) = c_2^{-1} \Phi_{c_1}^2(x) + \Phi_{c_2}^2(\psi_{c_1}(x))$$

a.e. $\mu(dx)$ for all $c_1, c_2 > 0$. By (2.98), the collection of maps $\{\Phi_c^2\}_{c>0}$ is an almost 1-semi-additive functional for the flow $\{\psi_c\}_{c>0}$ on (X,μ). It follows by Proposition 2.8 that $\{\Phi_c^2\}_{c>0}$ has a version $\{g_c\}_{c>0}$ which is a 1-semi-additive functional for the flow $\{\psi_c\}_{c>0}$.

As for the functions $\Psi_c(x)$, we also have by (3.36) that

$$\Psi_{c_1 c_2}(x) = \Psi_{c_1}(x) \Psi_{c_2}(\psi_{c_1}(x))$$

a.e. $\mu(dx)$. In other words, the collection $\{\Psi_c\}_{c>0}$ is an almost cocycle for the flow $\{\psi_c\}_{c>0}$ on (X,μ). Then, by Theorem 2.4, $\{\Psi_c\}_{c>0}$ has a version $\{b_c\}_{c>0}$ which is a cocycle for $\{\psi_c\}_{c>0}$. It takes values in $\{-1,1\}$ as stated in Definition 2.5.

Finally, consider the functions $\Phi_c^3(x)$. It is shown in Pipiras and Taqqu [41], Lemma 3.1, that the Jacobian

$$\frac{d(\mu \circ \psi_c)}{d\mu}$$

has a version which is a cocycle taking values in $\mathbb{R}$. Then, by Proposition 2.8, $\{\Phi_c^3\}_{c>0}$ has a version $\{j_c\}_{c>0}$ which is a 2-semi-additive functional for the flow $\{\psi_c\}_{c>0}$. $\square$

In the next result, we relate the flows and their cocycles and semi-additive functionals corresponding to *different* minimal representations of the same process. For simplicity and as in the relation (3.27), we include only 1-semi-additive functionals. A flow ψ, its related cocycle b, and 1-semi-additive functional g will be referred to as a *triplet* and denoted (ψ, b, g).

Definition 3.3. Triplets $(\psi^{(1)}, b^{(1)}, g^{(1)})$ and $(\psi^{(2)}, b^{(2)}, g^{(2)})$, where $b^{(i)} = \{b_c^{(i)}\}_{c>0}$ is a cocycle and $g^{(i)} = \{g_c^{(i)}\}_{c>0}$ is a 1-semi-additive functional for a measurable nonsingular multiplicative flow $\psi^{(i)} = \{\psi_c^{(i)}\}_{c>0}$ on (X_i, μ_i), $i = 1,2$, are said to be *equivalent*, denoted by

$$(\psi^{(1)}, b^{(1)}, g^{(1)}) \sim (\psi^{(2)}, b^{(2)}, g^{(2)}),$$

if there exists a measurable map $\Phi : X_2 \to X_1$ such that

(i) there is $N_i \subset X_i$ with $\mu_i(N_i) = 0$, $i = 1,2$, such that Φ is one-to-one, onto and bimeasurable between $X_2 \setminus N_2$ and $X_1 \setminus N_1$,

(ii) μ_1 and $\mu_2 \circ \Phi^{-1}$ are mutually absolutely continuous,

(iii) *relation between flows*: for all $c > 0$,

$$\psi_c^{(1)} \circ \Phi = \Phi \circ \psi_c^{(2)} \quad \text{a.e. } d\mu_2,$$

that is, the flows $\psi_c^{(1)}$ and $\psi_c^{(2)}$ are *a.e. isomorphic*,

(iv) *relation between cocycles*: the cocycle $\{b_c^{(1)} \circ \Phi\}_{c>0}$ is cohomologous to $\{b_c^{(2)}\}_{c>0}$, that is, there is a measurable function $b : X_2 \to \{-1,1\}$ such that, for all $c > 0$,

$$b_c^{(1)} \circ \Phi = b_c^{(2)} \cdot \frac{b \circ \psi_c^{(2)}}{b} \quad \text{a.e. } d\mu_2,$$

(v) *relation between semi-additive functionals*: there is a measurable function $g : X_2 \to \mathbb{R}$ such that

$$g_c^{(1)} \circ \Phi = g_c^{(2)} + g \circ \psi_c^{(2)} - c^{-1} g \quad \text{a.e. } d\mu_2.$$

Theorem 3.2. *Let*

$$\{G^{(i)}(x_i, t+u) - G^{(i)}(x_i, u), x_i \in X_i, u \in \mathbb{R}\}_{t \in \mathbb{R}} \subset L^\alpha(X_i \times \mathbb{R}, \mu_i(dx_i)du)$$

be two minimal representations for a SαS, self-similar mixed moving average X_α. Let also $(\psi^{(i)}, b^{(i)}, g^{(i)})$, $i = 1,2$, be the triplets corresponding to these minimal spectral representations by Theorem 3.1. Then,

$$(\psi^{(1)}, b^{(1)}, g^{(1)}) \sim (\psi^{(2)}, b^{(2)}, g^{(2)}).$$

PROOF: By using Proposition 3.5, there are a null-isomorphism $\Phi : X_2 \to X_1$ and functions

$$h : X_2 \to \mathbb{R} \setminus \{0\},\ g : X_2 \to \mathbb{R},\ j : X_2 \to \mathbb{R} \quad \text{and} \quad b : X_2 \to \{-1,1\}$$

such that

$$G^{(2)}(x_2, u) = h(x_2) G^{(1)}(\Phi(x_2), u + g(x_2)) + j(x_2), \tag{3.42}$$

$$h(x_2) = b(x_2) \left\{ \frac{d(\mu_1 \circ \Phi)}{d\mu_2}(x_2) \right\}^{1/\alpha} \tag{3.43}$$

a.e. $\mu_2(dx_2)du$. On the other hand, we know from Theorem 3.1 and Definition 3.2 that, for $i = 1,2$, and all $c > 0$,

$$c^{-\kappa} G^{(i)}(x_i, cu) = b_c^{(i)}(x_i) \left\{ \frac{d(\mu_i \circ \psi_c^{(i)})}{d\mu_i}(x_i) \right\}^{1/\alpha} G^{(i)}(\psi_c^{(i)}(x_i), u + g_c^{(i)}(x_i)) + j_c^{(i)}(x_i) \tag{3.44}$$

a.e. $\mu_i(dx_i)du$. The idea is to go from $G^{(2)}$ to $G^{(1)}$ using (3.42), then relating $G^{(1)}$ and $G^{(1)}$ using (3.44) with $i = 1$, and then returning to $G^{(2)}$ using (3.42) again and finally identifying the coefficients.

We have, by (3.42), for all $c > 0$,

$$c^{-\kappa} G^{(2)}(x_2, cu) = c^{-\kappa} h(x_2) G^{(1)}(\Phi(x_2), cu + g(x_2)) + c^{-\kappa} j(x_2).$$

By using (3.44) with $i=1$, where x_1 is replaced by $\Phi(x_2)$ and u by $u+c^{-1}g(x_2)$, the last expression becomes

$$h(x_2)b_c^{(1)}(\Phi(x_2))\left\{\frac{d(\mu_1\circ\psi_c^{(1)})}{d\mu_1}(\Phi(x_2))\right\}^{1/\alpha}\times$$
$$\times G^{(1)}(\psi_c^{(1)}(\Phi(x_2)),u+c^{-1}g(x_2)+g_c^{(1)}(\Phi(x_2)))$$
$$+h(x_2)j_c^{(1)}(\Phi(x))+c^{-\kappa}j(x_2).$$

By using (3.42) again, where x_2 is replaced by $(\Phi^{-1}\circ\psi_c^{(1)}\circ\Phi)(x_2)$ and then u by

$$u+c^{-1}g(x_2)+(g_c^{(1)}\circ\Phi)(x_2)-(g\circ\Phi^{-1}\circ\psi_c^{(1)}\circ\Phi)(x_2),$$

we can write $G^{(1)}$ in the last expression as

$$G^{(1)}(\psi_c^{(1)}(\Phi(x_2)),u+c^{-1}g(x_2)+g_c^{(1)}(\Phi(x_2)))$$
$$=G^{(1)}\Big(\Phi(\Phi^{-1}(\psi_c^{(1)}(\Phi(x_2)))),u+c^{-1}g(x_2)+g_c^{(1)}(\Phi(x_2))$$
$$-g(\Phi^{-1}(\psi_c^{(1)}(\Phi(x_2))))+g(\Phi^{-1}(\psi_c^{(1)}(\Phi(x_2))))\Big)$$
$$=h(\Phi^{-1}(\psi_c^{(1)}(\Phi(x_2))))^{-1}\times$$
$$\times G^{(2)}\Big(\Phi^{-1}(\psi_c^{(1)}(\Phi(x_2))),u+c^{-1}g(x_2)+g_c^{(1)}(\Phi(x_2))-g(\Phi^{-1}(\psi_c^{(1)}(\Phi(x_2))))\Big)$$
$$-h(\Phi^{-1}(\psi_c^{(1)}(\Phi(x_2))))^{-1}j(\Phi^{-1}(\psi_c^{(1)}(\Phi(x_2))))$$

and hence by substituting this, we can deduce that

$$c^{-\kappa}G^{(2)}(x_2,cu)=h(x_2)(b_c^{(1)}\circ\Phi)(x_2)\times$$
$$\times\left\{\Big(\frac{d\mu_1\circ\psi_c^{(1)}}{d\mu_1}\circ\Phi\Big)(x_2)\right\}^{1/\alpha}\Big((h\circ\Phi^{-1}\circ\psi_c^{(1)}\circ\Phi)(x_2)\Big)^{-1}\times$$
$$\times G^{(2)}((\Phi^{-1}\circ\psi_c^{(1)}\circ\Phi)(x_2),u+c^{-1}g(x_2)+(g_c^{(1)}\circ\Phi)(x_2)-(g\circ\Phi^{-1}\circ\psi_c^{(1)}\circ\Phi)(x_2))$$
$$-h(x_2)(b_c^{(1)}\circ\Phi)(x_2)\left\{\Big(\frac{d\mu_1\circ\psi_c^{(1)}}{d\mu_1}\circ\Phi\Big)(x_2)\right\}^{1/\alpha}\times$$
$$\times\Big((h\circ\Phi^{-1}\circ\psi_c^{(1)}\circ\Phi)(x_2)\Big)^{-1}(j\circ\Phi^{-1}\circ\psi_c^{(1)}\circ\Phi)(x_2)$$
$$+h(x_2)j_c^{(1)}(\Phi(x))+c^{-\kappa}j(x_2) \tag{3.45}$$

a.e. $\mu_2(dx_2)du$. By comparing (3.45) to (3.44) with $i=2$ and by using the uniqueness in Theorem 3.1, we conclude that for all $c>0$,

$$\psi_c^{(2)} = \Phi^{-1}\circ\psi_c^{(1)}\circ\Phi \quad \text{a.e. } d\mu_2, \tag{3.46}$$

$$b_c^{(1)}\circ\Phi = \frac{b\circ\psi_c^{(2)}}{b}b_c^{(2)} \quad \text{a.e. } d\mu_2, \tag{3.47}$$

$$g_c^{(1)}\circ\Phi = g_c^{(2)} + g\circ\psi_c^{(2)} - \frac{g}{c} \quad \text{a.e. } d\mu_2. \tag{3.48}$$

To obtain the equality (3.47), we also used (3.43) and

$$\left(\frac{d\mu_1\circ\Phi}{d\mu_2}\right)\left(\frac{d\mu_1\circ\psi_c^{(1)}}{d\mu_1}\circ\Phi\right)\left(\frac{d\mu_1\circ\Phi}{d\mu_2}\circ\Phi^{-1}\circ\psi_c^{(1)}\circ\Phi\right)^{-1}$$

$$= \frac{d\mu_2\circ\psi_c^{(2)}}{d\mu_2} \quad \text{a.e. } d\mu_2. \tag{3.49}$$

To verify (3.49), note that

$$\left(\frac{d\mu_1\circ\Phi}{d\mu_2}\right)\left(\frac{d\mu_1\circ\psi_c^{(1)}\circ\Phi}{d\mu_1\circ\Phi}\right) = \frac{d\mu_1\circ\psi_c^{(1)}\circ\Phi}{d\mu_2}$$

$$= \left(\frac{d\mu_1\circ\psi_c^{(1)}\circ\Phi}{d\mu_2\circ\Phi^{-1}\circ\psi_c^{(1)}\circ\Phi}\right)\left(\frac{d\mu_2\circ\psi_c^{(2)}}{d\mu_2}\right).$$

Since $\Phi^{-1}\circ\psi_c^{(1)}\circ\Phi = \psi_c^{(2)}$ by (3.46), we have indeed the equality after "cancelation" of measures in the numerators and denominators. Then, in view of relations (3.46)–(3.48) and Definition 3.3, we have $(\psi^{(1)}, b^{(1)}, g^{(1)}) \sim (\psi^{(2)}, b^{(2)}, g^{(2)})$.
□

3.2.2 Decomposition Related to Dissipative and Conservative Flows

Theorem 3.2 and part *(iii)* of Definition 3.3 show that the flows corresponding to two different minimal representations of a self-similar mixed moving average are "a.e. isomorphic." If two flows are a.e. isomorphic and one of them is dissipative (conservative, resp.), then the other will be the same. Based on this observation, one may then classify self-similar mixed moving averages into those whose minimal representations are associated with a dissipative flow and into those whose minimal representations are associated with a conservative flow.

In practice, however, it is not easy to say when a representation is minimal and therefore to determine whether a process is generated by a dissipative or a conservative flow. Since minimal representation kernels are of the form (3.25) or (3.27), it is best to use (not necessarily minimal) kernels of the form (3.25) as a starting point to derive the properties of the corresponding processes. In fact, many commonly used kernels (as in Examples 3.5–3.7) are already of the form (3.25).

The following result shows that the dissipative or conservative character of a flow is an invariant.

Theorem 3.3. *If the self-similar mixed moving average X_α, $\alpha \in (0,2)$, given by (3.1), (3.25), and (3.26), is generated by a dissipative (conservative, resp.) flow, then in any other representation (3.1), (3.25), and (3.26) of X_α, the multiplicative flow must be dissipative (conservative, resp.).*

PROOF: Suppose that the process X_α with the integral representation $\{G_t\}_{t\in\mathbb{R}}$ is generated by a multiplicative flow $\{\psi_c\}_{c>0}$ as in Definition 3.2. Let

$$X = D \cup C$$

be the Hopf decomposition of the flow $\{\psi_c\}_{c>0}$ (C and D denote the conservative and dissipative parts of the flow, respectively). Set also

$$F_t(x) = \int_{\mathbb{R}} |G_t(x,u)|^\alpha du = \int_{\mathbb{R}} |G(t,x+u) - G(x,u)|^\alpha du \tag{3.50}$$

and note that $F_t \in L^1(X,\mu)$ since $\int_X \int_{\mathbb{R}} |G_t(x,u)|^\alpha \mu(dx)du < \infty$ is a requirement for the stable process X_α to be defined. We will show that

$$D = \left\{ x \in X : \int_0^\infty \int_{\mathbb{R}} |G(\psi_c(x), 1+u) - G(\psi_c(x),u)|^\alpha du \, \lambda_c(x) \, c^{-1} dc < \infty \right\} \tag{3.51}$$

$$= \left\{ x \in X : \int_0^\infty F_1(\psi_c(x)) \, \lambda_c(x) \, c^{-1} dc < \infty \right\} \tag{3.52}$$

and

$$C = \left\{ x \in X : \int_0^\infty \int_{\mathbb{R}} |G(\psi_c(x), 1+u) - G(\psi_c(x),u)|^\alpha du \, \lambda_c(x) \, c^{-1} dc = \infty \right\} \tag{3.53}$$

$$= \left\{ x \in X : \int_0^\infty F_1(\psi_c(x)) \, \lambda_c(x) \, c^{-1} dc = \infty \right\} \tag{3.54}$$

a.e. $\mu(dx)$, where

$$\lambda_c = \frac{d(\mu \circ \psi_c)}{d\mu}. \tag{3.55}$$

Let D_0 and C_0 denote the right-hand side of (3.51) and (3.53), respectively. We want to show that $D = D_0$ and $C = C_0$ a.e. $\mu(dx)$, that is, D_0 and C_0 fully specify the dissipative part and conservative part, respectively.

In view of (2.80), by making the change of variables $c = e^v$ (which renders the flow additive), we have

$$D \subset D_0, \quad C \cap \mathrm{supp}\{F_1\} \subset C_0, \tag{3.56}$$

a.e. $\mu(dx)$. Since, for $t > 0$,

$$G(\psi_c(x), t+u) = G(\psi_c(x), t(1+t^{-1}u)),$$

we can apply (3.25) or (3.27). Observe first that

$$\lambda_t(\psi_c(x))\lambda_c(x) = \frac{d(\mu \circ \psi_t)}{d\mu}(\psi_c(x))\frac{d(\mu \circ \psi_c)}{d\mu}(x)$$

$$= \frac{d(\mu \circ \psi_t \circ \psi_c)}{d\mu}(x) = \frac{d(\mu \circ \psi_{ct})}{d\mu}(x) = \lambda_{ct}(x)$$

and also in (3.25), $|b_c(x)| = 1$ and the term $j_c(x)$ disappears once we take a difference of G's in (3.50). The relation (3.27) yields

$$\int_0^\infty F_t(\psi_c(x))\, \lambda_c(x)\, c^{-1} dc$$

$$= \int_0^\infty \int_{\mathbb{R}} |G(\psi_c(x), t(1+t^{-1}u)) - G(\psi_c(x), u)|^\alpha du\, \lambda_c(x)\, c^{-1} dc$$

$$= t^{\kappa\alpha} \int_0^\infty \int_{\mathbb{R}} \Big| G\Big(\psi_t(\psi_c(x)), 1+t^{-1}u + g_t(\psi_c(x))\Big)$$

$$- G\Big(\psi_t(\psi_c(x)), t^{-1}u + g_t(\psi_c(x))\Big)\Big|^\alpha du\, \lambda_t(\psi_c(x))\lambda_c(x)\, c^{-1} dc$$

$$= t^{\kappa\alpha} \int_0^\infty \int_{\mathbb{R}} |G(\psi_{ct}(x), 1+u) - G(\psi_{ct}(x), u)|^\alpha du\, t\lambda_{ct}(x)\, c^{-1} dc \tag{3.57}$$

$$= t^{\kappa\alpha+1} \int_0^\infty \int_{\mathbb{R}} |G(\psi_c(x), 1+u) - G(\psi_c(x), u)|^\alpha du\, \lambda_c(x)\, c^{-1} dc = t^{\kappa\alpha+1} F_1(x). \tag{3.58}$$

To get (3.57), we changed $t^{-1}u + g_t(\psi_c(x))$ into u. The factor $t^{\kappa\alpha}$ is due to the factor $c^{-\kappa}$ in (3.25). We can therefore replace F_1 by F_t, $t > 0$, in (3.54) and (3.52). This is also true for $t < 0$, since, by making the change of variables $t + u = v$ below,

$$\int_0^\infty \int_{\mathbb{R}} |G(\psi_c(x), t+u) - G(\psi_c(x), u)|^\alpha du\, \lambda_c(x)\, c^{-1} dc$$

$$= \int_0^\infty \int_{\mathbb{R}} |G(\psi_c(x), -t+v) - G(\psi_c(x), v)|^\alpha dv\, \lambda_c(x)\, c^{-1} dc.$$

Hence, in (3.56), we get that

$$C \cap \operatorname{supp}\{F_t\} \subset C_0 \text{ a.e. } \mu(dx), \quad \text{for all } t \in \mathbb{R}.$$

Since condition (3.26) implies that $\operatorname{supp}\{F_t, t \in \mathbb{R}\} = X$ a.e. $\mu(dx)$, it follows that $C \subset C_0$ a.e. $\mu(dx)$. Since X is a disjoint union of D and C and since D_0 and C_0 are disjoint, we get $D = D_0$ and $C = C_0$ a.e. $\mu(dx)$.

We will now show that the dissipative and conservative character of a flow is an invariant. By Proposition 3.2, the process X_α has also a minimal spectral representation $\{\widetilde{G}_t\}_{t\in\mathbb{R}}$ on the space $(\widetilde{X}\times\mathbb{R},\widetilde{\mu}(d\widetilde{x})du)$ and, by Theorem 3.1, it is generated by a multiplicative flow $\{\widetilde{\psi}_c\}_{c>0}$ on $(\widetilde{X},\widetilde{\mathscr{X}})$ associated with the kernel $\widetilde{G}$. We focus on (3.58). By using (3.25), Proposition 3.4 (with functions Φ_1 and h below) and the relation $-H\alpha=-\kappa\alpha-1$, we have a.e. $\mu(dx)$,

$$\int_0^\infty\int_{\mathbb{R}}|G(\psi_c(x),1+u)-G(\psi_c(x),u)|^\alpha du\,\lambda_c(x)\,c^{-1}dc$$

$$=\int_0^\infty\int_{\mathbb{R}}|G(\psi_c(x),1+u)-G(\psi_c(x),u)|^\alpha du\,\frac{d(\mu\circ\psi_c)}{d\mu}(x)\,c^{-1}dc$$

$$=\int_0^\infty\int_{\mathbb{R}}|G(x,c(1+u))-G(x,cu)|^\alpha du\,c^{-H\alpha}dc \tag{3.59}$$

$$=|h(x)|^\alpha\int_0^\infty\int_{\mathbb{R}}|\widetilde{G}(\Phi_1(x),c(1+u))-\widetilde{G}(\Phi_1(x),cu)|^\alpha du\,c^{-H\alpha}dc$$

$$=|h(x)|^\alpha\int_0^\infty\int_{\mathbb{R}}|\widetilde{G}(\widetilde{\psi}_c(\Phi_1(x)),1+u)-\widetilde{G}(\widetilde{\psi}_c(\Phi_1(x)),u)|^\alpha du\,\times$$

$$\times\,\frac{d(\widetilde{\mu}\circ\widetilde{\psi}_c)}{d\widetilde{\mu}}(\Phi_1(x))\,c^{-1}dc. \tag{3.60}$$

(Observe that the last equality in (3.60) holds a.e. $\mu(dx)$, since it holds a.e. $\widetilde{\mu}(d\widetilde{x})$ with $\widetilde{x}=\Phi_1(x)$ and since, by (3.20), $\widetilde{\mu}(\widetilde{N})=0$ for any $\widetilde{N}\in\widetilde{\mathscr{X}}$ implies that $\mu(\Phi_1^{-1}(\widetilde{N}))=0$.) This resembles the expression in (3.51), after setting

$$\widetilde{\lambda}_c(x)=\frac{d(\widetilde{\mu}\circ\widetilde{\psi}_c)}{d\widetilde{\mu}}(\Phi_1(x))$$

as in (3.55). Relation (3.60) implies that

$$\Phi_1^{-1}(\widetilde{D}_0)=D_0\quad\text{and}\quad\Phi_1^{-1}(\widetilde{C}_0)=C_0 \tag{3.61}$$

a.e. $\mu(dx)$, where D_0 and C_0 are the sets on the right-hand side of (3.51) and (3.53), respectively, and $\widetilde{D}_0$ and $\widetilde{C}_0$ are defined in the same way by replacing X, G, ψ_c, λ_c by $\widetilde{X}$, $\widetilde{G}$, $\widetilde{\psi}_c$, $\widetilde{\lambda}_c$. Hence, by (3.61), the flow $\{\psi_c\}_{c>0}$ is dissipative (conservative, resp.) if and only if the flow $\{\widetilde{\psi}_c\}_{c>0}$ is dissipative (conservative, resp.). Indeed, for example, if the flow $\{\psi_c\}_{c>0}$ is dissipative, then $\mu(C)=0$ and hence $\mu(C_0)=0$, which implies by (3.61) that $\mu(\Phi_1^{-1}(\widetilde{C}_0))=0$ and, by (3.20), that $\widetilde{\mu}(\widetilde{C}_0)=0$ or $\widetilde{\mu}(\widetilde{C})=0$. Now, if the process X_α is generated by yet another flow $\{\widehat{\psi}_c\}_{c>0}$ associated with a kernel $\widehat{G}$, then we conclude as above that the flow $\{\widehat{\psi}_c\}_{c>0}$ is dissipative (conservative, resp.) if and only if the flow $\{\widetilde{\psi}_c\}_{c>0}$ is dissipative (conservative, resp.), and consequently, if and only if the flow $\{\psi_c\}_{c>0}$ is dissipative (conservative, resp.). This concludes the proof. □

The next result provides a criterion for determining whether a flow is dissipative or conservative.

Theorem 3.4. *Let $\alpha \in (0,2)$ and X_α be a $S\alpha S$ H-self-similar process with stationary increments given by (3.1). Suppose that*

$$\operatorname{supp}\{G_t, t \in \mathbb{R}\} = X \times \mathbb{R}.$$

Then the process X_α is generated by a dissipative (conservative, resp.) flow in the sense of Definition 3.2 (with possibly a new kernel $\widetilde{G}$) if and only if the integral

$$I(x) = \int_0^\infty c^{-H\alpha} \int_{\mathbb{R}} |G(x, c(1+u)) - G(x, cu)|^\alpha \, du \, dc \tag{3.62}$$

is finite (infinite, resp.) a.e. $\mu(dx)$.

PROOF: By Proposition 3.2, the process X_α has a minimal spectral representation with the kernel function $\widetilde{G}$ and, by Theorem 3.1, there is a flow $\{\widetilde{\psi}_c\}_{c>0}$ associated with $\widetilde{G}$ which generates X_α in the sense of Definition 3.2. By Theorem 3.3, X_α is generated by a dissipative (conservative, resp.) flow if and only if the flow $\{\widetilde{\psi}_c\}_{c>0}$ is dissipative (conservative, resp.) and, by Proposition 3.4, the a.e finiteness of the integral (3.62) is equivalent to the a.e. finiteness of a similar integral where G is replaced by $\widetilde{G}$. By applying (3.25) with $\widetilde{G}$, we see as in (3.59) that the integral (3.62) with $\widetilde{G}$ equals to the integral

$$\int_0^\infty \int_{\mathbb{R}} |\widetilde{G}(\widetilde{\psi}_c(\widetilde{x}), 1+u) - \widetilde{G}(\widetilde{\psi}_c(\widetilde{x}), u)|^\alpha du \, \frac{d(\widetilde{\mu} \circ \widetilde{\psi}_c)}{d\widetilde{\mu}}(\widetilde{x}) \, c^{-1} dc.$$

The conclusion then follows from (3.51) ((3.53), resp.) with G, ψ_c, and D (C, resp.) replaced by $\widetilde{G}$, $\widetilde{\psi}_c$ and $\widetilde{D}$ ($\widetilde{C}$, resp.). □

Theorem 3.5. *Let $\alpha \in (0,2)$ and suppose that a process X_α is generated by a non-singular multiplicative flow $\{\psi_c\}_{c>0}$ as in Definition 3.2. Let also $X = D \cup C$ be the Hopf decomposition of the flow $\{\psi_c\}_{c>0}$. Then, we have*

$$X_\alpha \stackrel{d}{=} X_\alpha^D + X_\alpha^C, \tag{3.63}$$

where

$$X_\alpha^D(t) = \int_D \int_{\mathbb{R}} G_t(x,u) M(dx,du), \tag{3.64}$$

$$X_\alpha^C(t) = \int_C \int_{\mathbb{R}} G_t(x,u) M(dx,du). \tag{3.65}$$

The processes X_α^D and X_α^C are independent, and are both H-self-similar and have stationary increments. The process X_α^D is generated by a dissipative flow and the process X_α^C is generated by a conservative flow in the sense of Definition 3.2. The decomposition (3.63), moreover, is unique in distribution, that is, it does not depend on the representation $\{G_t\}_{t\in\mathbb{R}}$ in Definition 3.2.

PROOF: The processes X_α^D and X_α^C are independent because their kernels have disjoint support (Theorem 3.5.3 in Samorodnitsky and Taqqu [56]). The process X_α^D is generated by a dissipative flow and the process X_α^C is generated by a conservative flow in the sense of Definition 3.2 because D and C are invariant under the flow.

To prove the uniqueness in distribution, let $\{\widetilde{G}_t\}_{t\in\mathbb{R}} \subset L^\alpha(\widetilde{X}\times\mathbb{R}, \widetilde{\mu}(dx)du)$ be the minimal spectral representation of the process X_α obtained in Proposition 3.2. Suppose that this representation is generated by a multiplicative flow $\{\widetilde{\psi}_c\}_{c>0}$ on $(\widetilde{X}, \widetilde{\mu})$ as in Theorem 3.1. Let $\widetilde{D}$ and $\widetilde{C}$ be the dissipative part and the conservative parts of the flow $\{\widetilde{\psi}_c\}_{c>0}$, respectively. The kernels G and $\widetilde{G}$ can be related as in (3.19) and (3.20). Moreover, by (3.61), $\Phi_1^{-1}(\widetilde{D}_0) = D_0$ and $\Phi_1^{-1}(\widetilde{C}_0) = C_0$, where the sets D_0, C_0, $\widetilde{D}_0$, and $\widetilde{C}_0$ are defined in the proof of Theorem 3.3. Then, since $C = C_0$ μ-a.e. and $\widetilde{C} = \widetilde{C}_0$ $\widetilde{\mu}$-a.e., we have for every $\theta_1,\dots,\theta_n \in \mathbb{R}$ and $t_1,\dots,t_n \in \mathbb{R}$, $n \geq 1$, by (3.17),

$$
\begin{aligned}
\int_C \int_{\mathbb{R}} \Big| \sum_{k=1}^n \theta_k G_{t_k}(x,u) \Big|^\alpha \mu(dx)du &= \int_{C_0} \int_{\mathbb{R}} \Big| \sum_{k=1}^n \theta_k G_{t_k}(x,u) \Big|^\alpha \mu(dx)du \\
&= \int_{C_0} \int_{\mathbb{R}} \Big| \sum_{k=1}^n \theta_k \widetilde{G}_{t_k}(\Phi_1(x), u+\Phi_2(x)) \Big|^\alpha |h(x)|^\alpha \mu(dx)du \\
&= \int_{\Phi_1^{-1}(\widetilde{C}_0)} \int_{\mathbb{R}} \Big| \sum_{k=1}^n \theta_k \widetilde{G}_{t_k}(\Phi_1(x), u) \Big|^\alpha |h(x)|^\alpha \mu(dx)du \\
= \int_{\widetilde{C}_0} \int_{\mathbb{R}} \Big| \sum_{k=1}^n \theta_k \widetilde{G}_{t_k}(\widetilde{x},u) \Big|^\alpha \widetilde{\mu}(d\widetilde{x})du &= \int_{\widetilde{C}} \int_{\mathbb{R}} \Big| \sum_{k=1}^n \theta_k \widetilde{G}_{t_k}(\widetilde{x},u) \Big|^\alpha \widetilde{\mu}(d\widetilde{x})du. \quad (3.66)
\end{aligned}
$$

This implies that $X_\alpha^C \stackrel{d}{=} X_\alpha^{\widetilde{C}}$, where $X_\alpha^{\widetilde{C}}$ is defined analogously to X_α^C. Similarly, $X_\alpha^D \stackrel{d}{=} X_\alpha^{\widetilde{D}}$. It follows that the decomposition (3.63) does not depend on the representation $\{G_t\}_{t\in\mathbb{R}}$. □

Corollary 3.1. *When $\alpha \in (0,2)$, every H-self-similar process X_α having the representation (3.1) can be uniquely decomposed into two independent processes: one generated by a dissipative flow and the other generated by a conservative flow.*

PROOF: By Proposition 3.2 and Theorem 3.1, every H-self-similar process X_α having the representation (3.1) is generated by a multiplicative flow in the sense of Definition 3.2 (with possibly a new kernel $\widetilde{G}$). Then apply Theorem 3.5. □

We shall call the two processes obtained in Corollary 3.1 the *dissipative and conservative components* of the process X_α. Observe that they are defined in distribution. An alternative way to obtain the decomposition of the process X_α into its dissipative and conservative components is as follows.

Corollary 3.2. *Let $\alpha \in (0,2)$ and suppose that the H-self-similar process X_α has the representation (3.1) with* $\mathrm{supp}\{G_t, t\in\mathbb{R}\} = X\times\mathbb{R}$. *Define the sets*

$$
D = \{x \in X : I(x) < \infty\} \quad \textit{and} \quad C = \{x \in X : I(x) = \infty\}, \tag{3.67}
$$

where I is the integral defined in (3.62), and define the process X_α^D and X_α^C as in (3.64) and (3.65) but using the sets in (3.67). Then the processes X_α^D and X_α^C are (in distribution) the dissipative and conservative components of the process X_α.

PROOF: It is enough to show that

$$X_\alpha^D \stackrel{d}{=} X_\alpha^{\widetilde{D}} \quad \text{and} \quad X_\alpha^C \stackrel{d}{=} X_\alpha^{\widetilde{C}}, \tag{3.68}$$

where the processes $X_\alpha^{\widetilde{D}}$ and $X_\alpha^{\widetilde{C}}$ are the dissipative and conservative components of X_α defined in the proof of the uniqueness in Theorem 3.5. That proof also shows that (3.68) holds as long as

$$D = \Phi_1^{-1}(\widetilde{D}_0) \quad \text{and} \quad C = \Phi_1^{-1}(\widetilde{C}_0) \tag{3.69}$$

μ-a.e. (compare with (3.61)). These last relations hold, since by applying (3.19) and (3.25) with $\widetilde{G}$, we get as in (3.59)–(3.60) that a.e. $\mu(dx)$,

$$\begin{aligned} I(x) &= \int_0^\infty c^{-H\alpha} \int_{\mathbb{R}} |G(x,c(1+u)) - G(x,cu)|^\alpha \, du \, dc \\ &= |h(x)|^\alpha \int_0^\infty c^{-H\alpha} \int_{\mathbb{R}} \left|\widetilde{G}(\Phi_1(x),c(1+u)) - \widetilde{G}(\Phi_1(x),cu)\right|^\alpha du \, dc \\ &= |h(x)|^\alpha \int_0^\infty \int_{\mathbb{R}} |\widetilde{G}(\widetilde{\psi}_c(\Phi_1(x)),1+u) - \widetilde{G}(\widetilde{\psi}_c(\Phi_1(x)),u)|^\alpha du \times \\ &\quad \times \frac{d(\widetilde{\mu}\circ\widetilde{\psi}_c)}{d\widetilde{\mu}}(\Phi_1(x)) \; c^{-1} dc. \qquad \square \end{aligned}$$

Remark 3.6. The sets D and C in (3.67) are not related to a flow. But as indicated in the proof of the corollary, the processes X_α^D and X_α^C have the same distributions as $X_\alpha^{\widetilde{D}}$ and $X_\alpha^{\widetilde{C}}$, where $\widetilde{D}$ and $\widetilde{C}$ are the dissipative and conservative parts of a flow. This flow, which can be (but is not necessarily) associated with the minimal spectral representation of the process X_α, may live on a space which is different from X.

Example 3.8. LFSM in Example 3.5 is generated by a fixed flow, which is conservative. Note also that for this example, the function $I(x)$ in (3.62) is

$$\begin{aligned} I(1) &= \int_0^\infty c^{-H\alpha} \int_{\mathbb{R}} |G(1,c(1+u)) - G(1,cu)|^\alpha \, du \, dc \\ &= \int_0^\infty c^{-1} dc \int_{\mathbb{R}} |G(1,1+u) - G(1,u)|^\alpha \, du = \infty, \end{aligned}$$

where we used $G(1,cu) = c^\kappa G(1,u)$.

Example 3.9. The Telecom process in Example 3.6 is generated by the flow

$$\psi_c(x) = c^{-1}x \quad \text{on } x > 0.$$

The flow is dissipative since, for any $g \in L^1(0,\infty)$,

$$\int_0^\infty g(\psi_c(x)) \frac{d(\mathbb{L} \circ \psi_c)}{d\mathbb{L}}(x) \frac{dc}{c} = \int_0^\infty g(c^{-1}x) \frac{dc}{c^2} = x^{-1} \int_0^\infty g(w) dw < \infty,$$

after the change of variables $c^{-1}x = w$ (see (2.78)–(2.79)). Since by Example 3.6,

$$G(x, cu) = c^\kappa c^{-1/\alpha} G(c^{-1}x, u),$$

the function $I(x)$ in (3.62) is

$$\begin{aligned} I(x) &= \int_0^\infty c^{-H\alpha} \int_{\mathbb{R}} |G(x, c(1+u)) - G(x, cu)|^\alpha du\, dc \\ &= \int_0^\infty \int_{\mathbb{R}} \left|G(c^{-1}x, 1+u) - G(c^{-1}x, u)\right|^\alpha c^{-2} du dc \\ &= x \int_0^\infty \int_{\mathbb{R}} |G(w, 1+u) - G(w, u)|^\alpha du dw = x \|G_1\|_\alpha^\alpha < \infty, \end{aligned}$$

where we used the change of variables $c^{-1}x = w$.

Finally, the following definition introduces natural names for processes generated by dissipative and conservative flows that will be used below.

Definition 3.4. A $S\alpha S$ self-similar mixed moving average X_α is called

- *dissipative fractional stable motion (DFSM)* if it is generated by a dissipative flow, that is, if $X_\alpha \stackrel{d}{=} X_\alpha^D$,
- *conservative fractional stable motion (CFSM)* if it is generated by a conservative flow, that is, if $X_\alpha \stackrel{d}{=} X_\alpha^C$,

where X_α^D and X_α^C are the two components in the decomposition (3.63) of X_α obtained through a minimal representation.

In Sections 3.2.3 and 3.2.4 below, we take a closer look at DFSMs. In Sections 3.2.5, 3.2.6, and 3.2.8, we provide further decomposition of CFSMs and examine its components.

3.2.3 Canonical Representations of Processes Related to Dissipative Flows

In the next theorem, we show that $S\alpha S$ processes generated by dissipative multiplicative flows, that is, DFSMs, have a canonical representation. This canonical representation is very useful because it allows us to distinguish between different DFSMs. To obtain this canonical representation, we will be using Krengel's Theorem 2.3 in its multiplicative version.

Theorem 3.6. *Let $\alpha \in (0,2)$, $H > 0$, and*

$$\kappa = H - 1/\alpha.$$

Let also X_α be a $S\alpha S$ processes generated by a dissipative multiplicative flow as in Definition 3.2, that is, a DFSM. Then, there is a standard Lebesgue space $(Y, \mathscr{B}(Y), \nu)$ and a measurable function

$$F : Y \times \mathbb{R} \to \mathbb{R}$$

such that[5]

$$\{X_\alpha(t)\}_{t\in\mathbb{R}} \stackrel{d}{=} \left\{ \int_Y \int_{\mathbb{R}} \int_{\mathbb{R}} e^{-\kappa s}(F(y, e^s(t+u)) - F(y, e^s u)) M(dy, ds, du) \right\}_{t\in\mathbb{R}}, \tag{3.70}$$

where M is a $S\alpha S$ random measure on $Y \times \mathbb{R} \times \mathbb{R}$ with the control measure $m(dy, ds, du) = \nu(dy) ds du$. Conversely, if the process X_α has the representation (3.70), then it is a DFSM.

PROOF: According to Theorem 2.3, there is a standard Lebesgue space

$$(Y \times \mathbb{R}, \mathscr{B}(Y \times \mathbb{R}), \nu(dy)ds)$$

and a null-isomorphism

$$\Phi : Y \times \mathbb{R} \to X$$

such that

$$\psi_c(x) = \psi_c(\Phi(y,s)) = \Phi(y, s + \ln c), \tag{3.71}$$

for all $c > 0$ and $(y,s) \in Y \times \mathbb{R}$. In other words, the flow $\{\psi_c\}_{c>0}$ on $(X, \mathscr{B}(X), \mu)$ is null-isomorphic to a flow $\{\widetilde{\psi}_c\}_{c>0}$ on $(Y \times \mathbb{R}, \nu(dy)ds)$ defined by

$$\widetilde{\psi}_c(y,s) = (y, s + \ln c). \tag{3.72}$$

(We may suppose that the null sets in Theorem 2.3 are empty because, otherwise, we can replace X by $X \setminus N$ in the definition (3.1) of X_α without changing its distribution.) By replacing x by $\Phi(y,s)$ in (3.25) and using (3.71), we get for all $c > 0$

$$c^{-\kappa} G(\Phi(y,s), cu)$$

$$= b_c(\Phi(y,s)) \left\{ \frac{d(\mu \circ \psi_c)}{d\mu}(\Phi(y,s)) \right\}^{1/\alpha} G\Big(\Phi(y, s + \ln c), u + g_c(\Phi(y,s))\Big)$$
$$+ j_c(\Phi(y,s)) \tag{3.73}$$

[5]Do not confuse this function F with the one defined in (3.50).

a.e. $\nu(dy)dsdu$. One can show that $\{\widetilde{b}_c\}_{c>0} = \{b_c \circ \Phi\}_{c>0}$ is a cocycle, $\{\widetilde{g}_c\}_{c>0} = \{g_c \circ \Phi\}_{c>0}$ is a 1-semi-additive functional, and

$$\{\widetilde{j}_c\}_{c>0} = \Big\{\{\frac{d(\mu\circ\Phi)}{d(\nu\times\mathbb{L})}\}^{1/\alpha}(j_c\circ\Phi)\Big\}_{c>0}$$

is a 2-semi-additive functional for the multiplicative flow $\{\widetilde{\psi}_c\}_{c>0}$. By (2.92), (2.104), and (2.106) in Examples 2.14 and 2.17, they can be expressed as

$$b_c(\Phi(y,s)) = \frac{\widetilde{b}(y,s+\ln c)}{\widetilde{b}(y,s)}, \quad g_c(\Phi(y,s)) = \widetilde{g}(y,s+\ln c) - \frac{\widetilde{g}(y,s)}{c}, \tag{3.74}$$

$$\Big\{\frac{d(\mu\circ\Phi)}{d(\nu\times\mathbb{L})}(y,s)\Big\}^{1/\alpha} j_c(\Phi(y,s)) = \frac{\widetilde{b}(y,s+\ln c)}{\widetilde{b}(y,s)}\widetilde{j}(y,s+\ln c) - c^{-\kappa}\widetilde{j}(y,s), \tag{3.75}$$

for some measurable functions $\widetilde{b}$, taking values in $\{-1,1\}$, and $\widetilde{g}$, $\widetilde{j}$. Moreover, we have that ($\mathbb{L}$ denotes the Lebesgue measure)

$$\frac{d(\mu\circ\psi_c)}{d\mu}(\Phi(y,s)) = \frac{d(\mu\circ\Phi)}{d(\nu\times\mathbb{L})}(\widetilde{\psi}_c(y,s))\Big\{\frac{d(\mu\circ\Phi)}{d(\nu\times\mathbb{L})}(y,s)\Big\}^{-1}. \tag{3.76}$$

To show (3.76), observe first that (3.71) and (3.72) imply $\psi_c\circ\Phi = \Phi\circ\widetilde{\psi}_c$. Hence, we have

$$\frac{d(\mu\circ\psi_c\circ\Phi)}{d(\nu\times\mathbb{L})} = \frac{d(\mu\circ\Phi\circ\widetilde{\psi}_c)}{d(\nu\times\mathbb{L})}. \tag{3.77}$$

Relation (3.76) is merely a different expression for (3.77), since on one hand,

$$\frac{d(\mu\circ\psi_c\circ\Phi)}{d(\nu\times\mathbb{L})} = \frac{d(\mu\circ\psi_c\circ\Phi)}{d(\mu\circ\Phi)}\frac{d(\mu\circ\Phi)}{d(\nu\times\mathbb{L})} = \frac{d(\mu\circ\psi_c)}{d\mu}\circ\Phi\,\frac{d(\mu\circ\Phi)}{d(\nu\times\mathbb{L})},$$

and on the other hand,

$$\frac{d(\mu\circ\Phi\circ\widetilde{\psi}_c)}{d(\nu\times\mathbb{L})} = \frac{d(\nu\times\mathbb{L})\circ\widetilde{\psi}_c}{d(\nu\times\mathbb{L})}\frac{d(\mu\circ\Phi\circ\widetilde{\psi}_c)}{d(\nu\times\mathbb{L})\circ\widetilde{\psi}_c} = \frac{d(\mu\circ\Phi)}{d(\nu\times\mathbb{L})}\circ\widetilde{\psi}_c,$$

where

$$\frac{d((\nu\times\mathbb{L})\circ\widetilde{\psi}_c)}{d(\nu\times\mathbb{L})} = 1$$

because the first component in

$$\widetilde{\psi}_c(y,s) = (y,s+\ln c)$$

remains the same and the second is a translation. Now, by setting

$$\widetilde{G}(y,s,u) = \widetilde{b}(y,s)\Big(\Big\{\frac{d\mu}{d(\nu\times\mathbb{L})\circ\Phi^{-1}}(\Phi(y,s))\Big\}^{1/\alpha}G(\Phi(y,s),u) + \widetilde{j}(y,s)\Big) \tag{3.78}$$

using (3.74), (3.75), and (3.76), and after the change of variables $u = v + \widetilde{g}(y,s)/c$, the relation (3.73) can be expressed as, for all $c > 0$,

$$c^{-\kappa}\widetilde{G}(y,s,cv) = \widetilde{G}(y,s+\ln c, v)$$

a.e. $\nu(dy)dsdv$. By making the change of variables $cv = z$, we get

$$\widetilde{G}(y,s,z) = c^{\kappa}\widetilde{G}\Big(y,s+\ln c,\frac{z}{c}\Big)$$

a.e. $\nu(dy)dsdz$. By Lemma 1.1, this relation holds a.e $\nu(dy)dsdzdc$ as well. Then, by setting $u = s+\ln c$, we get $c = e^{u-s}$ and

$$\widetilde{G}(y,s,z) = e^{\kappa(u-s)}\widetilde{G}(y,u,e^{s-u}z)$$

a.e. $\nu(dy)dsdzdu$. Fix $u = u_0$ for which this relation holds a.e. $\nu(dy)dsdz$ and include the multiplicative terms $e^{\kappa u_0}$, e^{-u_0} in the function $\widetilde{G}$ on the right-hand side by renaming it to F. Then,

$$\widetilde{G}(y,s,z) = e^{-\kappa s}F(y,e^{s}z), \tag{3.79}$$

a.e. $\nu(dy)dsdz$ for some measurable function F. Now, by using (3.78), for any $\theta_1,\dots,\theta_n \in \mathbb{R}$, $t_1,\dots,t_n \in \mathbb{R}$,

$$\int_X\int_{\mathbb{R}}\Big|\sum_{k=1}^{n}\theta_k(G(x,t_k+u)-G(x,u))\Big|^{\alpha}\mu(dx)du$$

$$= \int_Y\int_{\mathbb{R}}\int_{\mathbb{R}}\Big|\sum_{k=1}^{n}\theta_k(G(\Phi(y,s),t_k+u)-G(\Phi(y,s),u))\Big|^{\alpha}d\mu(\Phi(y,s))du$$

$$= \int_Y\int_{\mathbb{R}}\int_{\mathbb{R}}\Big|\sum_{k=1}^{n}\theta_k(G(\Phi(y,s),t_k+u)-G(\Phi(y,s),u))\Big|^{\alpha}\frac{d(\mu\circ\Phi)}{d(\nu\times\mathbb{L})}(y,s)\nu(dy)dsdu$$

$$= \int_Y\int_{\mathbb{R}}\int_{\mathbb{R}}\Big|\sum_{k=1}^{n}\theta_k(\widetilde{G}(y,s,t_k+u)-\widetilde{G}(y,s,u))\Big|^{\alpha}\nu(dy)dsdu,$$

since $|\widetilde{b}(y,s)| = 1$ in (3.78) and the shift $\widetilde{b}(y,s)\widetilde{j}(y,s)$ in (3.78) cancels out by taking the difference. This yields

$$\{X_\alpha(t)\}_{t\in\mathbb{R}} \overset{d}{=} \Big\{\int_X\int_{\mathbb{R}}(G(x,t+u)-G(x,u))M(dx,du)\Big\}_{t\in\mathbb{R}}$$

$$\overset{d}{=} \Big\{\int_Y\int_{\mathbb{R}}\int_{\mathbb{R}}(\widetilde{G}(y,s,t+u)-\widetilde{G}(y,s,u))\widetilde{M}(dy,ds,du)\Big\}_{t\in\mathbb{R}},$$

where $\widetilde{M}$ is a $S\alpha S$ random measure on $Y\times\mathbb{R}\times\mathbb{R}$ with the control measure $\widetilde{m}(dy,ds,du) = \nu(dy)dsdu$. Then, by using (3.79), we get that

$$\{X_\alpha(t)\}_{t\in\mathbb{R}} \overset{d}{=} \Big\{\int_Y\int_{\mathbb{R}}\int_{\mathbb{R}} e^{-\kappa s}\Big(F(y,e^{s}(t+u))-F(y,e^{s}u)\Big)\widetilde{M}(dy,ds,du)\Big\}_{t\in\mathbb{R}}.$$

To prove the converse, suppose that the process X_α has the representation (3.70). By using an argument similar to the one in the proof of Lemma 4.2 in Pipiras and Taqqu [35], one can conclude that there is a set $Y_0 \in \mathscr{Y}$ such that

$$\mathrm{supp}\{e^{-\kappa s}(F(y,e^s(t+u))-F(y,e^s u)),\ t\in\mathbb{R}\} = Y_0\times\mathbb{R}\times\mathbb{R} \quad \text{a.e. } \nu(dy)dsdu.$$

Hence, by replacing Y with Y_0 in (3.70), we may suppose that the condition (3.26) holds. We may do this without loss of generality because this replacement does not change the distribution of X_α. We shall now show that relation (3.25) is satisfied with

$$G(y,s,u) = e^{-\kappa s}F(y,e^s u),$$

where the x in this relation stands for (y,s). Observe that since $c^{-\kappa}G(y,s,cu) = G(y,s+\ln c,u)$ for any $c>0$, the condition (3.25) is satisfied with

$$\psi_c(y,s) = (y,s+\ln c),\ b_c(y,s)=1,\ g_c(y,s)=0 \quad \text{and} \quad j_c(y,s)=0.$$

The collection $\{\psi_c\}_{c>0}$ is clearly a dissipative multiplicative flow. □

There are many other ways to represent processes (3.70) as indicated in the following lemma.

Lemma 3.1. *Other representations for the process X_α in (3.70) are as follows:*

$$X_\alpha(t) \overset{d}{=} \int_Y\int_0^\infty\int_{\mathbb{R}} z^{-H}(F(y,z(t+u))-F(y,zu))M(dy,dz,du) \tag{3.80}$$

$$\overset{d}{=} \int_Y\int_0^\infty\int_{\mathbb{R}} z^{H-\frac{2}{\alpha}}(F(y,z^{-1}(t+u))-F(y,z^{-1}u))M(dy,dz,du) \tag{3.81}$$

$$\overset{d}{=} \int_Y\int_{\mathbb{R}}\int_{\mathbb{R}} e^{-Hs}(F(y,e^s t+v)-F(y,v))M(dy,ds,dv) \tag{3.82}$$

$$\overset{d}{=} \int_Y\int_0^\infty\int_{\mathbb{R}} z^{-H-\frac{1}{\alpha}}(F(y,zt+v)-F(y,v))M(dy,dz,dv) \tag{3.83}$$

$$\overset{d}{=} \int_Y\int_0^\infty\int_{\mathbb{R}} z^{H-\frac{1}{\alpha}}(F(y,z^{-1}t+v)-F(y,v))M(dy,dz,dv), \tag{3.84}$$

where $\overset{d}{=}$ denotes the equality in the sense of the finite-dimensional distributions. The control measure of M is $\nu(dy)dzdu$.

PROOF: The lemma follows by making proper changes of variables in (3.70). □

Further study of DFSMs can be found in Pipiras and Taqqu [37],[6] including the study of integrand spaces F, uniqueness questions, and examples. We include some of those examples below.

---

[6]In Pipiras and Taqqu [37], dissipative fractional stable motions (DFSMs) are called dilated fractional stable motions.

Example 3.10. We know from Examples 3.2, 3.9, and 3.6 that the Telecom process is generated by a dissipative flow. By writing its kernel function

$$G(x,u) = (u\wedge 0+x)_+ x^{H-\frac{2}{\alpha}-1} = ((x^{-1}u)\wedge 0+1)_+ x^{H-\frac{2}{\alpha}} =: F(x^{-1}u)x^{H-\frac{2}{\alpha}},$$

note that the Telecom process has a canonical representation (3.81) with

$$F(1,z) = (z\wedge 0+1)_+, \quad z\in\mathbb{R},$$

and $Y=\{1\}$, $\nu(dy)=\delta_{\{1\}}(dy)$.

Example 3.11. Consider $Y=\{1\}$, $\nu(dy)=\delta_{\{1\}}(dy)$ and the function

$$F(1,z) = 1_{\{|z|\le 1\}}(z), \quad z\in\mathbb{R},$$

in the canonical representation (3.70), corresponding to the process

$$\begin{aligned} X_\alpha(t) &= \int_{\mathbb{R}}\int_{\mathbb{R}} (1_{\{e^s|t+u|\le 1\}}(s,u) - 1_{\{e^s|u|\le 1\}}(s,u))\, e^{-(H-\frac{1}{\alpha})s}\, M(ds,du) \\ &\overset{d}{=} \int_0^\infty \int_{\mathbb{R}} (1_{\{|u-t|<x\}}(x,u) - 1_{\{|u|<x\}}(x,u))\, x^{H-\frac{2}{\alpha}}\, M(dx,du), \qquad (3.85) \end{aligned}$$

where we set $x=e^{-s}$. One can verify that the process (3.85) is well defined if $\alpha\in(0,2)$ and $H\in(0,1/\alpha)$.

This last expression resembles the H-self-similar process with stationary increments constructed by means of integral geometry in Takenaka [62]. If, for $t\in\mathbb{R}$, we let

$$A_t = \{(x,u)\in\mathbb{R}_+\times\mathbb{R} : |u-t|<x\}$$

and define

$$B_t = A_0\,\triangle\, A_t = (A_0\setminus A_t)\cup(A_t\setminus A_0),$$

then the process constructed by Takenaka [62] for $\alpha\in(0,2)$ and $H\in(0,1/\alpha)$ is given by

$$\begin{aligned} \widetilde{X}_\alpha(t) &= \int_0^\infty \int_{\mathbb{R}} 1_{B_t}(x,u) x^{H-\frac{2}{\alpha}} \widetilde{M}(dx,du) \\ &= \int_0^\infty \int_{\mathbb{R}} |1_{A_t}(x,u) - 1_{A_0}(x,u)| x^{H-\frac{2}{\alpha}} \widetilde{M}(dx,du) \\ &= \int_0^\infty \int_{\mathbb{R}} |1_{\{|u-t|<x\}}(x,u) - 1_{\{|u|<x\}}(x,u)| x^{H-\frac{2}{\alpha}} \widetilde{M}(dx,du), \end{aligned}$$

where $\widetilde{M}$ has the Lebesgue control measure (see also Samorodnitsky and Taqqu [56], Chapter 8).

The difference between X_α and $\widetilde{X}_\alpha$ is that the kernel in the definition of $\widetilde{X}_\alpha$ is taken in absolute value. Nevertheless, the processes X_α and $\widetilde{X}_\alpha$ have the same finite-dimensional distributions. This is a consequence of the following simple observation,

$$1_{\{|u-t|<x\}}(x,u) - 1_{\{|u|<x\}}(x,u) = h(x,u)|1_{\{|u-t|<x\}}(x,u) - 1_{\{|u|<x\}}(x,u)|,$$

for all t,u,x, where $h(x,u) = -1$, if $|u| < x$, and $h(x,u) = 1$, if $|u| \geq x$ (draw a picture in the (u,x) plane). Since h does not depend on t and since $|h(x,u)| = 1$, one can use characteristic functions to prove the equality of the finite-dimensional distributions. Therefore, the processes (3.85) are *Takenaka processes*.

Example 3.12. Let $\alpha \in (0,2)$, $H \in (0,1)$ and $a \in \mathbb{R}$ be such that either $H < 1/\alpha < H - a$ (here, $a < 0$) or $H - a < 1/\alpha < H$ (here, $a > 0$). Then, the process

$$X_\alpha(t) = \int_0^\infty \int_{\mathbb{R}} \left((t+u)_+^a \wedge x^a - u_+^a \wedge x^a\right) x^{H-\frac{2}{\alpha}-a} M(dx,du), \quad t \in \mathbb{R},$$

where M has the Lebesgue control measure, is called a *mixed truncated fractional stable motion*. To derive the canonical representation (3.70), write

$$\begin{aligned} X_\alpha(t) &\stackrel{d}{=} \int_{\mathbb{R}} \int_{\mathbb{R}} \left(e^{sa}(t+u)_+^a \wedge 1 - e^{sa}u_+^a \wedge 1\right) e^{-(H-\frac{1}{\alpha})s} M(ds,du) \\ &= \int_{\mathbb{R}} \int_{\mathbb{R}} e^{-\kappa s}(F(e^s(t+u)) - F(e^s u))\, M(ds,du), \end{aligned}$$

where $\kappa = H - 1/\alpha$ and

$$F(z) = z_+^a \wedge 1, \quad z \in \mathbb{R}.$$

(Note that the shape of the function F is quite different for $a > 0$ and $a < 0$.) For more information on the mixed truncated fractional stable motion, see Surgailis et al. [61].

Example 3.13. When $\alpha \in (0,2)$ and $H \in (0,1)$, Samorodnitsky and Taqqu [55] introduced a class of $S\alpha S$, H-self-similar processes with stationary increments given by

$$X_\alpha(t) = \int_{\mathbb{R}^n} \left(\|t\mathbf{1} + \mathbf{u}\|^{H-\frac{n}{\alpha}} - \|\mathbf{u}\|^{H-\frac{n}{\alpha}}\right) M(du), \quad t \in \mathbb{R}, \tag{3.86}$$

where $n \geq 2$, $\|\cdot\|$ is the usual Euclidean norm, $\mathbf{u} = (u_1,\ldots,u_n) \in \mathbb{R}^n$ and $\mathbf{1} = (1,\ldots,1) \in \mathbb{R}^n$, and a $S\alpha S$ random measure M has the Lebesgue control measure. Let us show that the processes (3.86) have the canonical representation (3.70). Consider the case $n = 2$ first. By making the change of variables $u_1 = u$ and $u_2 = u + v$, we can first represent (3.86) with $n = 2$ as the mixed moving average

$$X_\alpha(t) \stackrel{d}{=} \int_{\mathbb{R}^2} \left(((t+u)^2 + (t+u+v)^2)^{\frac{H}{2}-\frac{1}{\alpha}} - (u^2 + (u+v)^2)^{\frac{H}{2}-\frac{1}{\alpha}}\right) M(du,dv). \tag{3.87}$$

To see that (3.87) has the canonical representation (3.70), consider the integral in (3.87) over the regions $v > 0$ and $v < 0$ separately and make the changes of variables $v = e^{-s}$ and $v = -e^{-s}$, respectively. Then, by setting

$$F(1,z) = (z^2 + (z+1)^2)^{\frac{H}{2}-\frac{1}{\alpha}}, \quad F(2,z) = (z^2 + (z-1)^2)^{\frac{H}{2}-\frac{1}{\alpha}}, \quad z \in \mathbb{R}, \tag{3.88}$$

and $Y = \{1,2\}$, we have

$$X_\alpha(t) \stackrel{d}{=} \int_Y \int_{\mathbb{R}} \int_{\mathbb{R}} e^{-(H-\frac{1}{\alpha})s}(F(y,e^s(t+u)) - F(y,e^s u))\, M(dy,ds,du), \qquad (3.89)$$

where M has the control measure $\delta_{\{1,2\}}(dy)dsdu$, which is the representation (3.70).

In the case $n \geq 3$, one can represent the process (3.86) through the DFSM representation (3.89) with the space $Y = \{1,2\} \times \mathbb{R}^{n-2}$ and the corresponding points $y = (y_0, y_1, \dots, y_{n-2})$, where $y_0 \in \{1,2\}$ and $(y_1, \dots, y_{n-2}) \in \mathbb{R}^{n-2}$, the $S\alpha S$ random measure $M(dy,ds,du)$ with the control measure $\delta_{\{1,2\}}(dy_0)dy_1 \dots dy_{n-2}dsdu$ and the kernel function

$$F(1,y_1,\dots,y_{n-2},z) = \left(z^2 + (z+1)^2 + (z+y_1)^2 + \dots + (z+y_{n-2})^2\right)^{\frac{H}{2}-\frac{n}{2\alpha}},$$

$$F(2,y_1,\dots,y_{n-2},z) = \left(z^2 + (z-1)^2 + (z+y_1)^2 + \dots + (z+y_{n-2})^2\right)^{\frac{H}{2}-\frac{n}{2\alpha}}, \qquad (3.90)$$

where $z \in \mathbb{R}$.

We shall call the processes X_α in (3.86) *Samorodnitsky processes*. Observe that these processes provide examples of $1/\alpha$-self-similar processes with stationary increments by setting $H = 1/\alpha$ and $\alpha \in (1,2)$ (so that $H \in (0,1)$). Samorodnitsky and Taqqu showed that these processes are different from the usual log-fractional stable motion and the independent increments stable Lévy motion.

It can be shown (Pipiras and Taqqu [37]) that a self-similar mixed moving average given by a canonical representation (3.70) is well defined if and only if

$$\int_Y \int_{\mathbb{R}} \int_{\mathbb{R}} \frac{|F(y,z_1) - F(y,z_2)|^\alpha}{|z_1 - z_2|^{\alpha H+1}}\, \nu(dy)dz_1dz_2 < \infty.$$

3.2.4 On the Uniqueness of Processes Related to Dissipative Flows

One would like to know whether DFSMs, which are defined by (3.70), are really different for different F's, namely whether they have different finite-dimensional distributions. The following result, analogous to Proposition 3.3, can often be used to conclude that the distributions of two given DFSMs are different.

Proposition 3.6. *Let $\alpha \in (0,2)$ and $H > 0$. Suppose that the process X_α has two representations (3.70): one on the standard Lebesgue space (Y, ν) and with the function $F : Y \times \mathbb{R} \to \mathbb{R}$, and the other on the standard Lebesgue space $(\widetilde{Y}, \widetilde{\nu})$ and with the function $\widetilde{F} : \widetilde{Y} \times \mathbb{R} \to \mathbb{R}$. Then, there exist measurable functions $\Phi_1 : Y \to \widetilde{Y}$ and $h, \Phi_2, \Phi_3, \Phi_4 : Y \to \mathbb{R}$ such that*

$$F(y,z) = h(y)\widetilde{F}(\Phi_1(y), e^{\Phi_2(y)}(z + \Phi_3(y))) + \Phi_4(y) \qquad (3.91)$$

a.e. $\nu(dy)dz$.

PROOF: By Proposition 2.5, there exist measurable functions $\Phi_1 : Y \times \mathbb{R} \times \mathbb{R} \to \widetilde{Y}$ and $h, \Phi_2, \Phi_3 : Y \times \mathbb{R} \times \mathbb{R} \to \mathbb{R}$ such that

$$e^{-\kappa s}(F(y,e^s(t+u)) - F(y,e^s u)) = h(y,s,u)e^{-\kappa\Phi_2(y,s,u)} \times$$
$$\times \Big(\widetilde{F}(\Phi_1(y,s,u), e^{\Phi_2(y,s,u)}(t+u+\Phi_3(y,s,u)))$$
$$-\widetilde{F}(\Phi_1(y,s,u), e^{\Phi_2(y,s,u)}(u+\Phi_3(y,s,u)))\Big)$$

a.e. $\nu(dy)dsdudt$. By making the change of variables $t = e^{-s}z - u$, we get that

$$F(y,z) = F(y,e^s u) + e^{\kappa s}h(y,s,u)e^{-\kappa\Phi_2(y,s,u)} \times$$
$$\times \Big(\widetilde{F}(\Phi_1(y,s,u), e^{\Phi_2(y,s,u)}(e^{-s}z+\Phi_3(y,s,u)))$$
$$-\widetilde{F}(\Phi_1(y,s,u), e^{\Phi_2(y,s,u)}(u+\Phi_3(y,s,u)))\Big)$$

a.e. $\nu(dy)dsdudz$. By fixing $s = s_0$ and $u = u_0$, for which this equation holds a.e. $\nu(dy)dz$ and setting $h(y) = e^{\kappa s_0}h(y,s_0,u_0)e^{-\kappa\Phi_2(y,s_0,u_0)}$, we obtain the result. □

Observe that Proposition 3.6 does not provide an "if and only if" condition. One can obtain, however, such a condition in the special case $Y = \widetilde{Y} = \{1\}$ and $\nu = \widetilde{\nu} = \delta_{\{1\}}$.

Corollary 3.3. *Suppose that the processes X_α and $\widetilde{X}_\alpha$ have the representation (3.70) with the spaces $Y = \widetilde{Y} = \{1\}$, the measures $\nu = \widetilde{\nu} = \delta_{\{1\}}$ and the functions F and $\widetilde{F}$, respectively. If X_α and $\widetilde{X}_\alpha$ have the same finite-dimensional distributions, then*

$$F(z) = a\widetilde{F}(bz+c)+d \quad \text{a.e. } dz, \text{ for some constants } a,c,d \in \mathbb{R} \text{ and } b > 0. \tag{3.92}$$

Conversely, if condition (3.92) holds, then the processes X_α and $\widetilde{X}_\alpha$ have the same finite-dimensional distributions up to a multiplicative constant. They have identical finite-dimensional distributions if $|a|b^\kappa = 1$.

PROOF: The first statement of the corollary follows from Proposition 3.6. To show the second statement, assume that the condition (3.92) holds and let $t_k, \theta_k \in \mathbb{R}$, $k = 1,\dots,n$. Then, by making the change of variables $u = \tilde{u} - b^{-1}e^{-s}c$ and $s = \tilde{s} - \ln b$ below, we have

$$-\ln \mathbb{E}\exp\Big\{i\sum_{k=1}^n \theta_k X_\alpha(t_k)\Big\} = \int_{\mathbb{R}}\int_{\mathbb{R}}\Big|\sum_{k=1}^n \theta_k e^{-\kappa s}(F(e^s(t_k+u)) - F(e^s u))\Big|^\alpha dsdu$$
$$= \int_{\mathbb{R}}\int_{\mathbb{R}}\Big|\sum_{k=1}^n \theta_k e^{-\kappa s}a(\widetilde{F}(be^s(t_k+u)+c) - \widetilde{F}(be^s u+c))\Big|^\alpha dsdu$$
$$= (|a|b^\kappa)^\alpha \int_{\mathbb{R}}\int_{\mathbb{R}}\Big|\sum_{k=1}^n \theta_k e^{-\kappa\tilde{s}}(\widetilde{F}(e^{\tilde{s}}(t_k+\tilde{u})) - \widetilde{F}(e^{\tilde{s}}\tilde{u}))\Big|^\alpha d\tilde{s}d\tilde{u}$$

$$= -\ln \mathbb{E} \exp\Big\{ i \sum_{k=1}^{n} \theta_k |a| b^{\kappa} \widetilde{X}_\alpha(t_k) \Big\}.$$

This show that the processes X_α and $|a|b^\kappa \widetilde{X}_\alpha$ have the same finite-dimensional distributions. □

In practice, one may not want to distinguish between processes which differ by a multiplicative constant. Hence, we say that two processes X and $\widetilde{X}$ are *essentially identical* if there exists a multiplicative constant c such that $X(t)$ and $c\widetilde{X}(t)$ have the same finite-dimensional distributions. If these processes are not essentially identical, we say that they are *essentially different*. It is easy to see that the following holds.

Corollary 3.4. *(i) Suppose that X_α has the representation (3.70) with the standard Lebesgue space (Y,ν) and with the function $F: Y \times \mathbb{R} \to \mathbb{R}$, and $\widetilde{X}_\alpha$ has the representation (3.70) with the standard Lebesgue space $(\widetilde{Y},\widetilde{\nu})$ and with the function $\widetilde{F}: \widetilde{Y} \times \mathbb{R} \to \mathbb{R}$. If condition (3.91) does not hold, then the processes X_α and $\widetilde{X}_\alpha$ are essentially different.*

(ii) Suppose that the assumptions of Corollary 3.3 hold. Then the processes X_α and $\widetilde{X}_\alpha$ are essentially identical if and only if condition (3.92) holds.

The second part of this corollary shows that the Telecom process in Example 3.10, defined by the function

$$F_1(z) = (z \wedge 0 + 1)_+,\ z \in \mathbb{R},$$

the Takenaka process in Example 3.11, defined by the function

$$F_2(z) = 1_{\{|z| \le 1\}},\ z \in \mathbb{R},$$

and the mixed truncated fractional stable motion in Example 3.12, defined by the function

$$F_3(z) = z_+^a \wedge 1,\ z \in \mathbb{R},$$

(if $a \neq 1$) are essentially different. In the case $a = 1$, $F_1(z) = F_3(z+1)$ and therefore the Telecom process in Example 3.10 is essentially identical to a mixed truncated fractional stable motion in Example 3.12. Since $|a|b^\kappa = 1$, the two process have, in fact, the same finite-dimensional distributions.

By using Proposition 3.6, one may also show that the Samorodnitsky processes in Example 3.13 are essentially different from the processes in Examples 3.10, 3.11, and 3.12. Indeed, let

$$F(y,z) = F(i, y_1, \ldots, y_{n-2}, z)$$

be the kernel function (3.90) of the process X_α in Example 3.13, where

$$y = (i, y_1, \ldots, y_{n-2}) \in \{1,2\} \times \mathbb{R}^{n-2}$$

and $\widetilde{F}(z)$ be one of the kernel functions of processes $\widetilde{X}_\alpha$ in Examples 3.10, 3.11, and 3.12. If the processes X_α and $\widetilde{X}_\alpha$ are essentially identical, then by Proposition 3.6, there are functions $a, c, d: Y \to \mathbb{R}$ and $b: Y \to (0,\infty)$ such that

$$F(y,z) = a(y)\widetilde{F}(b(y)z + c(y)) + d(y) \tag{3.93}$$

a.e. $\nu(dy)dz$, and also by reversing the roles of the processes in Proposition 3.6, there are $\widetilde{y} \in Y$, $\widetilde{a}, \widetilde{c}, \widetilde{d} \in \mathbb{R}$ and $\widetilde{b} > 0$ such that

$$\widetilde{F}(z) = \widetilde{a}F(\widetilde{y}, \widetilde{b}z + \widetilde{c}) + \widetilde{d} \tag{3.94}$$

a.e. dz. It is easy to verify that neither the relation (3.93) nor the relation (3.94) holds. (In general, it is enough to verify that only one of the relations (3.93) and (3.94), whichever is easier to deal with, does not hold.) For example, if

$$\widetilde{F}(z) = z_+^a \wedge 1$$

is the kernel function in Example 3.12, then for $y = (1, y_1, \ldots, y_{n-2})$, the relation (3.93) becomes

$$\begin{aligned} &\left(z^2 + (z+1)^2 + (z+y_1)^2 + \cdots + (z+y_{n-2})^2\right)^{\frac{H}{2} - \frac{n}{2\alpha}} \\ &= a(1, y_1, \ldots, y_{n-2})\left(\left(b(1, y_1, \ldots, y_{n-2})z + c(1, y_1, \ldots, y_{n-2})\right)_+^a \wedge 1\right) \\ &\quad + d(1, y_1, \ldots, y_{n-2}) \end{aligned} \tag{3.95}$$

a.e. $dy_1 \ldots dy_{n-2}dz$, and the relation (3.94) becomes

$$z_+^a \wedge 1 = \widetilde{a}\left((\widetilde{b}z + \widetilde{c})^2 + (\widetilde{b}z + \widetilde{c} + (-1)^{\widetilde{i}+1})^2 + \cdots + (\widetilde{b}z + \widetilde{c} + \widetilde{y}_{n-2})^2\right)^{\frac{H}{2} - \frac{n}{2\alpha}} + \widetilde{d} \tag{3.96}$$

a.e. dz, for some fixed $\widetilde{y} = (\widetilde{i}, \widetilde{y}_1, \ldots, \widetilde{y}_{n-2})$ where $\widetilde{i} = 1, 2$. Neither the relation (3.95) nor the relation (3.96) holds. For example, by fixing $y_1, \ldots, y_{n-2}$ so that the relation (3.95) holds a.e. dz and taking z negative large enough, the right-hand side of (3.95) becomes a constant while the left-hand side behaves like a power of z. Hence, the processes X_α and $\widetilde{X}_\alpha$ are essentially different.

Since Samorodnitsky and Taqqu [55] have shown that the processes in Example 3.13 differ from each other if they correspond to different values of n, we get:

Proposition 3.7. *The processes in Examples 3.10, 3.11, 3.12 ($a \neq 1$), and 3.13 are essentially different.*

3.2.5 Further Decomposition of Processes Related to Conservative Flows

We now turn to processes generated by conservative flows, that is, CFSMs X_α^C in the decomposition (3.63). In contrast to Section 3.2.3, we will not provide a canonical representation of such processes because conservative flows cannot be characterized as simply as dissipative flows are by Krengel's Theorem 2.3. Instead, we will pursue a different idea, though some of the developments will shed further light on the results of previous sections.

By Proposition 3.2, any $S\alpha S$ mixed moving average X_α has an integral representation (3.1) which is minimal. By Theorem 3.1, a self-similar mixed moving averages X_α given by a minimal representation (3.1) is generated by a unique flow $\{\psi_c\}_{c>0}$ in the sense of Definition 3.2. As in Definition 2.3, let

$$D,\ C,\ F,\ L \text{ and } P$$

be the dissipative, conservative, fixed, cyclic, and periodic point sets of the flow $\{\psi_c\}_{c>0}$, respectively. Recall that $P = F + L$. Since

$$X = D + C = D + P + C\setminus P = D + F + L + C\setminus P$$

(see (2.89)), we can write[7]

$$X_\alpha \stackrel{d}{=} X_\alpha^D + X_\alpha^C = X_\alpha^D + X_\alpha^P + X_\alpha^{C\setminus P} = X_\alpha^D + X_\alpha^F + X_\alpha^L + X_\alpha^{C\setminus P}, \tag{3.97}$$

where

$$X_\alpha^P = X_\alpha^F + X_\alpha^L$$

and where, for a set $S \subset X$,

$$X_\alpha^S(t) = \int_S \int_{\mathbb{R}} G_t(x,u) M(dx,du). \tag{3.98}$$

Since by their definitions, the sets D, C, F, P, and L are invariant under the flow, the processes

$$X_\alpha^D,\ X_\alpha^F,\ X_\alpha^L \text{ and } X_\alpha^{C\setminus P}$$

are self-similar mixed moving averages. These processes are independent because the sets D, F, L, and $C\setminus P$ are disjoint (see Theorem 3.5.3 in Samorodnitsky and Taqqu [56]). The processes X_α^S are generated by the flow ψ^S where ψ^S denotes the flow ψ restricted to a set S, which is invariant under the flow. Observe that ψ^D, ψ^F, ψ^L, and ψ^P are dissipative, fixed, cyclic, and periodic flows, respectively, and that $\psi^{C\setminus P}$ is a conservative flow without periodic points, and, for example, the process X_α^D is generated by the dissipative flow ψ^D.

A self-similar mixed moving average may have another minimal representation (3.1) with a kernel function $\widetilde{G}$ on the space $\widetilde{X}$, and hence be generated by another flow $\{\widetilde{\psi}_c\}_{c>0}$. Partitioning $\widetilde{X}$ into the dissipative, fixed, cyclic, and conservative nonperiodic point sets of the flow $\{\widetilde{\psi}_c\}_{c>0}$ as above leads to the decomposition

$$X_\alpha \stackrel{d}{=} \widetilde{X}_\alpha^D + \widetilde{X}_\alpha^F + \widetilde{X}_\alpha^L + \widetilde{X}_\alpha^{C\setminus P}. \tag{3.99}$$

[7]Do not confuse the process X_α with the set X.

We will say that the decomposition (3.97) obtained from a minimal representation (3.1) is *unique in distribution* if the distribution of its components does not depend on the minimal representation used in the decomposition. In other words, uniqueness in distribution holds if

$$X_\alpha^D \stackrel{d}{=} \widetilde{X}_\alpha^D, \quad X_\alpha^F \stackrel{d}{=} \widetilde{X}_\alpha^F, \quad X_\alpha^L \stackrel{d}{=} \widetilde{X}_\alpha^L, \quad X_\alpha^{C\setminus P} \stackrel{d}{=} \widetilde{X}_\alpha^{C\setminus P}, \tag{3.100}$$

where X_α^S and $\widetilde{X}_\alpha^S$ with

$$S = D, F, L \text{ and } C\setminus P,$$

are the components of the decompositions (3.97) and (3.99) obtained from two different minimal representations of the process.

Theorem 3.7. *The decomposition (3.97) obtained from a minimal representation (3.1) of a self-similar mixed moving average X_α is unique in distribution.*

PROOF: In accordance with the notation above, let D,F,L,P,C and $\widetilde{D},\widetilde{F},\widetilde{L},\widetilde{P},\widetilde{C}$ be the dissipative, fixed, cyclic, periodic, and conservative point sets of the flows $\{\psi_c\}_{c>0}$ and $\{\widetilde{\psi}_c\}_{c>0}$, respectively. We need to show that the equalities (3.100) hold.

By Theorem 3.2 (see also (3.42) and (3.43)), there is a map $\Phi : \widetilde{X} \to X$ such that

(i) Φ is (mod 0) null-isomorphism;
(ii) $\widetilde{\mu}\circ\Phi$ and μ are mutually absolutely continuous;
(iii) for all $c>0$, $\psi_c \circ \Phi = \Phi \circ \widetilde{\psi}_c$ $\widetilde{\mu}$-a.e., and
(iv)

$$\widetilde{G}(\widetilde{x},u) = b(\widetilde{x})\left\{\frac{d(\mu\circ\Phi)}{d\widetilde{\mu}}(\widetilde{x})\right\}^{1/\alpha} G(\Phi(\widetilde{x}), u+g(\widetilde{x})) + j(\widetilde{x}), \quad \text{a.e. } \widetilde{\mu}(d\widetilde{x})du, \tag{3.101}$$

where $b:\widetilde{X}\to\{-1,1\}$ and $g,j:\widetilde{X}\to\mathbb{R}$ are measurable functions.

By (3.61),

$$\Phi^{-1}(D) = \widetilde{D}, \quad \Phi^{-1}(C) = \widetilde{C}, \quad \widetilde{\mu} - \text{a.e.} \tag{3.102}$$

By using relations (i)–(iii), we can deduce directly from (2.83) to (2.85) that

$$\Phi^{-1}(F) = \widetilde{F}, \quad \Phi^{-1}(P) = \widetilde{P}, \quad \Phi^{-1}(L) = \widetilde{L}, \quad \widetilde{\mu}\text{-a.e.} \tag{3.103}$$

and hence

$$\Phi^{-1}(C\setminus P) = \widetilde{C}\setminus\widetilde{P}, \quad \widetilde{\mu}\text{-a.e.} \tag{3.104}$$

The equalities (3.100) can now be obtained by using (3.101) together with (3.102)–(3.104) as, for example, in the proof of Theorem 3.5 (see (3.66)). □

Since the decomposition (3.97) can be obtained through a minimal representation for any $S\alpha S$ self-similar mixed moving average, and it is unique in distribution by Theorem 3.7, we may give the following definition.

Definition 3.5. A $S\alpha S$ self-similar mixed moving average X_α is called

- *periodic fractional stable motion* if $X_\alpha \stackrel{d}{=} X_\alpha^P$,
- *cyclic fractional stable motion* if $X_\alpha \stackrel{d}{=} X_\alpha^L$,
- *fixed fractional stable motion*[8] if $X_\alpha \stackrel{d}{=} X_\alpha^F$,
- *conservative nonperiodic fractional stable motion* if $X_\alpha \stackrel{d}{=} X_\alpha^{C\setminus P}$,

where X_α^F, X_α^L, X_α^P, and $X_\alpha^{C\setminus P}$ are the four components in the decomposition (3.97) of X_α obtained through a minimal representation.

Notation. Periodic, cyclic, fixed, and conservative nonperiodic fractional stable motion will be abbreviated as *PFSM*, *cLFSM*, *FFSM*, and *(C\P)FSM*, respectively.[9]

An equivalent definition of these motions is as follows.

Proposition 3.8. *A $S\alpha S$ self-similar mixed moving average is a periodic (cyclic, fixed, conservative nonperiodic resp.) fractional stable motion if and only if the generating flow corresponding to its minimal representation is periodic (cyclic, fixed, conservative nonperiodic resp.).*

PROOF: By Definition 3.5, a self-similar mixed moving average X_α is

a PFSM (cLFSM, FFSM, (C\P)FSM, resp.)

if and only if

$$X_\alpha \stackrel{d}{=} X_\alpha^P \; (X_\alpha \stackrel{d}{=} X_\alpha^L, X_\alpha \stackrel{d}{=} X_\alpha^F, X_\alpha \stackrel{d}{=} X_\alpha^{C\setminus P} \text{resp.}),$$

where P (L, resp.) is the set of periodic (cyclic, resp.) points of the generating flow ψ corresponding to a minimal representation. It follows from (3.97) and (3.98) that $X_\alpha \stackrel{d}{=} X_\alpha^P$ ($X_\alpha \stackrel{d}{=} X_\alpha^L$, $X_\alpha \stackrel{d}{=} X_\alpha^F$, $X_\alpha \stackrel{d}{=} X_\alpha^{C\setminus P}$, resp.) if and only if $X = P$ ($X = L, X = F$, $X = C \setminus P$, resp.) μ-a.e. and hence if and only if the flow ψ is periodic (cyclic, fixed, conservative nonperiodic, resp.). □

Definition 3.5 and Proposition 3.8 use minimal representations. Minimal representations, however, are not very easy to determine in practice. It is therefore desirable to recognize a PFSM and a cLFSM based on any, possibly nonminimal representation.

Since many self-similar mixed moving averages given by nonminimal representations are generated by a flow in the sense of Definition 3.2, we could expect that the process is a PFSM (cLFSM, resp.) if the generating flow is periodic (cyclic, resp.). This, however, is not the case in general. For example, if a PFSM or cLFSM

---

[8] In Pipiras and Taqqu [36] (see Definition 5.1), the fixed fractional stable motion was called mixed linear fractional stable motion because of Theorem 3.11 below.

[9] Cyclic fractional stable motion is abbreviated as cLFSM, since the abbreviation LFSM is commonly used for linear fractional stable motion.

$$X_\alpha(t) = \int_X \int_{\mathbb{R}} G_t(x,u) M(dx,du)$$

is generated by a periodic or cyclic flow $\psi_c(x)$ on X, we can also represent the process X_α as

$$\int_Y \int_X \int_{\mathbb{R}} G_t(x,u) M(dy,dx,du),$$

where $G_t(x,u)$ does not depend on y and the control measure $\eta(dy)$ of $M(dy,dx,du)$ in the variable y is such that $\eta(Y) = 1$. If $\widetilde{\psi}_c(y)$ is a measure preserving flow on (Y,η), then the process X_α is also generated by the flow

$$\Phi_c(y,x) = (\widetilde{\psi}_c(y), \psi_c(x)) \quad \text{on } Y \times X.$$

If, in addition, the flow $\widetilde{\psi}_c(y)$ is not periodic (and hence not cyclic), then the flow $\Phi_c(y,x)$ is neither periodic nor cyclic. Thus, a PFSM may be generated by a nonperiodic flow.

We will provide identification criteria for a PFSM, a cLFSM, a FFSM, and a (C\P)FSM which do not rely on either minimal representations or flows, and which are based instead on the structure of the kernel function G. We will focus on PFSMs and will only outline the identification of cLFSMs, FFSMs, and (C\P)FSMs. Thus, let X_α be a self-similar mixed moving average (3.1) defined through a (possibly nonminimal) kernel function G.

Definition 3.6. A *periodic fractional stable motion set* (*PFSM set*, in short) of a self-similar mixed moving average X_α given by (3.1) is defined as

$$C_P = \Big\{ x \in X : \exists\, c = c(x) \neq 1 : G(x,cu) = bG(x,u+a) + d \ \text{ a.e. } du \\ \text{for some } a = a(c,x), b = b(c,x) \neq 0, d = d(c,x) \in \mathbb{R} \Big\}. \tag{3.105}$$

Remark 3.7. The relation in (3.105) can be expressed as

$$G(x,cu+g) = bG(x,u+g) + d \tag{3.106}$$

for some $b \neq 0$, $c \neq 1$, $g,d \in \mathbb{R}$. When $b \neq 1$, it can also be expressed as

$$G(x,cu+g) + f = b(G(x,u+g) + f) \tag{3.107}$$

for some $b \neq 0$, $c \neq 1$, $g,f \in \mathbb{R}$. Relation (3.106) follows by making the change of variables $u = v + a/(c-1)$ in $G(x,cu) = bG(x,u+a) + d$ and setting $g = ca/(c-1)$. When $b \neq 1$, by writing $d = bf - f$ with $f = d/(b-1)$ in (3.106), we get (3.107).

Remark 3.8. C_P can be understood heuristically as follows: discounting a, b, and d, we have $x \in C_P$ if for almost every "time" u, there is a different "time" cu, where G takes again the same value, namely $G(x,cu) = G(x,u)$.

Whereas the set of periodic points P is defined by (2.83) in terms of the flow $\{\psi_c\}_{c>0}$, the set C_P in (3.105) is defined in terms of the kernel G. Definition 3.6 states that there is a factor c such that the kernel G at time u is related to the kernel at time cu.

Lemma 3.2. *The PFSM set C_P in (3.105) is μ-measurable. Moreover, the functions $c(x), a(x) = a(c(x),x)$, $b = b(c(x),x)$, and $d = d(c(x),x)$ in (3.105) can be taken to be μ-measurable as well.*

PROOF: We first show the measurability of C_L. Consider the set

$$A = \Big\{(x,c,a,b,d) : G(x,cu) = bG(x,u+a) + d \ \text{ a.e. } du\Big\}.$$

Since $A = \{F(x,c,a,b,d) = 0\}$, where the function

$$F(x,c,a,b,d) = \int_{\mathbb{R}} 1_{\{G(x,cu)=bG(x,u+a)+d\}}(x,c,a,b,d,u)du$$

is measurable by Fubini's theorem, we obtain that the set A is measurable. Observe that the set C_P is a projection of the set A on x, namely, that

$$C_P = \text{proj}_X A := \{x : \exists\, c,a,b,d : (x,c,a,b,d) \in A\}.$$

Lemma B.1 in Appendix B implies that the PFSM set C_P is μ-measurable and that the functions $a(x)$, $b(x)$, $c(x)$, and $d(x)$ can be taken to be μ-measurable as well. □

In the next theorem, we characterize a PFSM in terms of the set C_P instead of using the set P which involves flows as is done in Definition 3.6 and Proposition 3.8. Flows and minimal representations, however, are used in the proof.

Theorem 3.8. *A $S\alpha S$ self-similar mixed moving average X_α given by (3.1) with G satisfying (3.26) is a PFSM if and only if $C_P = X$ μ-a.e., where C_P is the PFSM set defined in (3.105).*

PROOF: Suppose first that X_α is a self-similar mixed moving average given by (3.1) with G satisfying (3.26) and such that $C_P = X$ μ-a.e. We need to show that X_α is a PFSM. The proof has 2 steps.

Step 1: We will show without loss of generality that the representation (3.1) can be supposed to be minimal with $C_P = X$ μ-a.e. By Proposition 3.2, the process X_α has a minimal integral representation

$$\int_{\widetilde{X}} \int_{\mathbb{R}} \Big(\widetilde{G}(\widetilde{x}, t+u) - \widetilde{G}(\widetilde{x}, u)\Big) \widetilde{M}(d\widetilde{x}, du), \tag{3.108}$$

where $(\widetilde{X}, \widetilde{\mathscr{X}}, \widetilde{\mu})$ is a standard Lebesgue space and $\widetilde{M}(d\widetilde{x}, du)$ has control measure $\widetilde{\mu}(d\widetilde{x})du$. Letting $\widetilde{C}_P$ be the periodic component set of X_α defined using the kernel function $\widetilde{G}$, we need to show that $\widetilde{C}_P = \widetilde{X}$ $\widetilde{\mu}$-a.e. By Proposition 3.4, there are measurable maps

$$\Phi_1 : X \to \widetilde{X},\ h : X \to \mathbb{R}\setminus\{0\} \quad \text{and} \quad \Phi_2, \Phi_3 : X \to \mathbb{R}$$

such that

$$G(x,u) = h(x)\widetilde{G}(\Phi_1(x), u+\Phi_2(x)) + \Phi_3(x) \tag{3.109}$$

a.e. $\mu(dx)du$, and

$$\widetilde{\mu} = \mu_h \circ \Phi_1^{-1}, \tag{3.110}$$

where

$$\mu_h(dx) = |h(x)|^\alpha \mu(dx).$$

If $x \in C_P$, then

$$G(x, c(x)u) = b(x)G(x, u+a(x)) + d(x) \quad \text{a.e. } du, \tag{3.111}$$

for some functions $a(x), b(x)$, $c(x)$, and $d(x)$. Hence, by using (3.109) and (3.111), we have for some functions F_1, F_2, and F_3, a.e. $\mu(dx)$,

$$\widetilde{G}(\Phi_1(x), c(x)u+\Phi_2(x)) = (h(x))^{-1}G(x, c(x)u) + F_1(x)$$

$$= (h(x))^{-1}b(x)G(x, u+a(x)) + F_2(x) = b(x)\widetilde{G}(\Phi_1(x), u+a(x)+\Phi_2(x)) + F_3(x)$$

a.e. du. This shows that $\Phi_1(x) \in \widetilde{C}_P$ and hence

$$C_P \subset \Phi_1^{-1}(\widetilde{C}_P), \quad \mu\text{-a.e.} \tag{3.112}$$

Since $C_P = X$ μ-a.e., we have $X = \Phi^{-1}(\widetilde{C}_P)$ μ-a.e. This implies $\widetilde{C}_P = \widetilde{X}$ $\widetilde{\mu}$-a.e., because if $\widetilde{\mu}(\widetilde{X} \setminus \widetilde{C}_P) > 0$, then by (3.110), we have

$$\mu(\Phi_1^{-1}(\widetilde{X} \setminus \widetilde{C}_P)) = \mu(\Phi_1^{-1}(\widetilde{X}) \setminus X) = \mu(\emptyset) > 0.$$

Remark 3.9. The converse is shown in the same way: if C_P is not equal to X μ-a.e., then $\Phi_1^{-1}(\widetilde{C}_P) \subset C_P$ μ-a.e. Together with (3.112), this implies

$$C_P = \Phi_1^{-1}(\widetilde{C}_P) \quad \mu\text{-a.e.} \tag{3.113}$$

The relation (3.113) is used in the proof of Theorem 3.10 below.

We may therefore suppose without loss of generality that the representation (3.1) is minimal and that $C_P = X$ μ-a.e. By Theorem 3.1, since the representation (3.1) is minimal, the process X_α is generated by a flow $\{\psi_c\}_{c>0}$ and related functionals $\{b_c\}_{c>0}$, $\{g_c\}_{c>0}$, and $\{j_c\}_{c>0}$ in the sense of Definition 3.2.

Step 2: To conclude the proof, it is enough to show, by Proposition 3.8, that the flow $\{\psi_c\}_{c>0}$ is periodic. The idea can informally be explained as follows. By using (3.25) and (3.105), we get that for $c = c(x) \neq 1$,

$$G(\psi_{c(x)}(x), u) = h(x)G(x, c(x)u+a(x)) + j(x) = k(x)G(x, u+b(x)) + l(x),$$

for some $a,b,h \neq 0, j,k \neq 0,l$. Then, for any $t \in \mathbb{R}$, after replacing G by G_t,

$$G_t(\Psi(x,u)) = k(x)G_t(x,u), \tag{3.114}$$

where

$$\Psi(x,u) = (\psi_{c(x)}(x), u - b(x)) \quad \text{and} \quad k(x) \neq 0.$$

Since the representation $\{G_t\}_{t\in\mathbb{R}}$ is minimal, (3.114) and (2.4)–(2.5) imply $\Psi(x,u) = (x,u)$ and therefore $\psi_{c(x)}(x) = x$ for $c(x) \neq 1$, showing that the flow $\{\psi_c\}_{c>0}$ is periodic. This argument is not rigorous because c depends on x and hence the relation (3.25) cannot be applied directly. A rigorous proof can be found in Theorem 4.1 of Pipiras and Taqqu [40].

To prove the converse, suppose that X_α given by (3.1) with a kernel G satisfying (3.26) is a PFSM. We need to show that $C_P = X$ μ-a.e. By Proposition 3.8, the minimal representation (3.108) of X_α is generated by a periodic flow $\{\widetilde{\psi}_c\}_{c>0}$. Let $\widetilde{P}$ be the set of the periodic points of the flow $\{\widetilde{\psi}_c\}_{c>0}$, and $\widetilde{C}_P$ be the PFSM set defined using the representation (3.108). Since the flow $\{\widetilde{\psi}_c\}_{c>0}$ is periodic, $\widetilde{P} = \widetilde{X}$ a.e. $\widetilde{\mu}(d\widetilde{x})$. Since $\widetilde{P} = \widetilde{C}_P$ a.e. $\widetilde{\mu}(d\widetilde{x})$ by Proposition 3.9 below, we have $\widetilde{C}_P = \widetilde{X}$ a.e. $\widetilde{\mu}(d\widetilde{x})$. In addition, the following three equalities hold a.e. $\mu(dx)$:

$$C_P = \Phi_1^{-1}(\widetilde{C}_P), \Phi_1^{-1}(\widetilde{C}_P) = \Phi_1^{-1}(\widetilde{X}) \text{ and } \Phi_1^{-1}(\widetilde{X}) = X.$$

The first equality follows from Step 1 and more specifically (3.113), the second holds because the measures $\mu \circ \Phi_1^{-1}$ and $\widetilde{\mu}$ are absolutely continuous by (3.110) and hence $\widetilde{C}_P = \widetilde{X}$ a.e. $\widetilde{\mu}(d\widetilde{x})$ implies

$$\mu(\Phi_1^{-1}(\widetilde{X} \setminus \widetilde{C}_P)) = 0.$$

The third equality follows from the definition of Φ_1. Stringing these equalities together one gets $C_P = X$ a.e. $\mu(dx)$. $\square$

The next result describes relations between the PFSM set C_P defined using a kernel function G, and the set of periodic points P of a flow related to the kernel G as in Definition 3.2. The first part of the result was used in the proof of Theorem 3.8 above.

Proposition 3.9. *Suppose that a SαS self-similar mixed moving average X_α given by (3.1) is generated by a flow $\{\psi_c\}_{c>0}$. Let P be the set of periodic points of the flow $\{\psi_c\}_{c>0}$ and C_P the PFSM set (3.105) defined using the kernel G of the representation (3.1). Then, we have*

$$P \subset C_P \quad \mu\text{-a.e.} \tag{3.115}$$

If, moreover, the representation (3.1) is minimal, we have

$$P = C_P \quad \mu\text{-a.e.} \tag{3.116}$$

PROOF: We first prove (3.115). Let $\tau(dc)$ denote any σ-finite measure on $(0,\infty)$. By Lemma 1.1, the relation (3.25) implies that a.e. $\mu(dx)\tau(dc)$,

$$G(x,cu) = hG(\psi_c(x),u+g)+j \quad \text{a.e. } du, \tag{3.117}$$

for some $h=h(x,c)\neq 0$, $g=g(x,c)$ and $j=j(x,c)$. Hence, setting

$$\widetilde{P} := \Big\{(x,c)\in X\times((0,\infty)\setminus\{1\}) : \psi_c(x)=x\Big\},$$

we have a.e. $\mu(dx)\tau(dc)$,

$$\widetilde{P} = \widetilde{P}\bigcap\Big\{(x,c) : G(x,cu) = hG(\psi_c(x),u+g)+j \text{ a.e. } du \text{ for some } h\neq 0,g,j\Big\}$$

$$= \widetilde{P}\bigcap\Big\{(x,c) : G(x,cu) = hG(x,u+g)+j \text{ a.e. } du \text{ for some } h\neq 0,g,j\Big\}. \tag{3.118}$$

Since $P=\mathrm{proj}_X\widetilde{P}$, the relation (3.118) implies that a.e. $x\in P$ belongs to the set

$$\mathrm{proj}_X\Big(\Big\{(x,c) : G(x,cu) = hG(x,u+g)+j \text{ a.e. } du \text{ for some } h\neq 0,g,j,c\neq 1\Big\}\Big),$$

that is, there is $c=c(x)\neq 1$ such that

$$G(x,cu) = hG(x,u+g)+j \quad \text{a.e. } du$$

for some $h\neq 0,g,j$. This shows that $P\subset C_P$ a.e. $\mu(dx)$.

To prove (3.116), suppose that the representation (3.1) is minimal. It is enough to show that $C_P\subset P$ μ-a.e. Let $\{G_t|_{C_P}\}$ be the kernel G_t of (3.1) restricted to the set $C_P\times\mathbb{R}$. By Lemma 3.3 below, the set C_P is a.e. invariant under the flow $\{\psi_c\}$. Then, $\{G_t|_{C_P}\}$ is a representation of a self-similar mixed moving average. Since $\{G_t\}$ is minimal, so is the representation $\{G_t|_{C_P}\}$. It is obviously generated by the flow $\psi|_{C_P}$, the restriction of the flow ψ to the set C_P. Therefore, for a.e. $x\in C_P$, $\psi_{c(x)}(x)=x$ for some $c(x)\neq 1$. One can show this as in Step 2 of the proof of Theorem 3.8. This shows that $C_P\subset P$ a.e. $\mu(dx)$. □

The following lemma was used in the proof of Proposition 3.9 above.

Lemma 3.3. *If a $S\alpha S$ self-similar mixed moving average X_α given by a representation (3.1) is generated by a flow $\{\psi_c\}_{c>0}$, and C_P is the PFSM set defined by (3.105), then C_P is a.e. invariant under the flow $\{\psi_c\}_{c>0}$, that is,*

$$\mu(C_P\triangle\psi_c^{-1}(C_P))=0 \quad \textit{for all } c>0.$$

PROOF: Since $\{\psi_c\}_{c>0}$ is a flow, it is enough to show that $C_P\subset\psi_r^{-1}(C_P)$ μ-a.e. for any fixed $r>0$. Indeed, since $\psi_{1/r}$ is the inverse map of ψ_r, $C_P\subset\psi_r^{-1}(C_P)$ is equivalent to $\psi_{1/r}^{-1}(C_P)\subset C_P$ and hence $1/r$ can be replaced by r in the last relation to get the reverse inclusion. By (3.25), we have for any $c>0$,

$$G(\psi_r(x), cu + a(x)) = b(x)G(x, cru) + j(x) \quad \text{a.e. } \mu(dx)du, \tag{3.119}$$

for some $a, b \neq 0, j$ (these depend on r but since r is fixed, we do not indicate their dependence on r).

We want to show next that there is a function $c(x) \neq 1$ such that for a.e. $x \in C_P$,

$$G(x, c(x)ru) = b(x)G(x, ru + a(x)) + j(x) \quad \text{a.e. } du, \tag{3.120}$$

for some $a, b \neq 0, j$, and such that at the same time, the relation (3.119) holds with c replaced by $c(x)$. Set

$$A = \Big\{(x,c) \in X \times (0,\infty) \setminus \{1\} : G(\psi_r(x), cu + a(x)) = b(x)G(x, cru) + j(x) \text{ a.e. } du$$
$$\text{for some } a(x), b(x) \neq 0, j(x)\Big\}$$

and

$$B = \Big\{(x,c) \in X \times (0,\infty) \setminus \{1\} : G(x, c(x)ru) = b(x)G(x, ru + a(x)) + j(x) \text{ a.e. } du$$
$$\text{for some } a(x), b(x) \neq 0, j(x)\Big\}.$$

By (3.105), $\text{proj}_X B = C_P$. By (3.119), for any $c > 0$, $1_A(x,c) = 1$ a.e. $\mu(dx)$ and hence by Lemma 1.1, $1_A(x,c) = 1$ a.e. $\mu(dx)dc$. Then,

$$1_{A\cap B}(x,c) = 1_B(x,c) \quad \text{a.e. } \mu(dx)dc$$

or $A \cap B = B$ a.e. $\mu(dx)dc$. In particular,

$$\text{proj}_X(A \cap B) = \text{proj}_X B = C_P \quad \text{a.e. } \mu(dx).$$

By using Lemma B.1 in Appendix B, there is indeed $c = c(x) \neq 1$ defined on $\text{proj}_X(A \cap B) = C_P$ a.e. $\mu(dx)$ such that (3.120) and (3.119) both hold with $c = c(x)$.

By substituting (3.120) into (3.119) with $c = c(x)$ and then making a change of variables in u to eliminate $a(x)$ on the right-hand side of (3.120), we obtain that for a.e. $x \in C_P$,

$$G(\psi_r(x), c(x)u + d(x)) = h(x)G(x, ru) + l(x) \quad \text{a.e. } du,$$

for some $d, h \neq 0, l$. Then, by using (3.25) and making a change of variables in u, we get that for a.e. $x \in C_P$,

$$G(\psi_r(x), c(x)u) = k(x)G(\psi_r(x), u + p(x)) + q(x) \quad \text{a.e. } du,$$

for some $k \neq 0, p, q$, indicating that $\psi_r(x) \in C_P$. Hence, for a.e. $x \in C_P$, we have $\psi_r(x) \in C_P$ or $x \in \psi_r^{-1}(C_P)$, showing that $C_P \subset \psi_r^{-1}(C_P)$ μ-a.e. $\square$

We introduced in Definition 3.5 the fixed fractional stable motion (FFSM), the cyclic fractional stable motion (cLFSM), the conservative nonperiodic fractional stable motion ((C\P)FSM), and the periodic fractional stable motion (PFSM). The PFSM identifying set C_P was given in (3.105). We indicate next, without providing proofs, the identifying sets C_F for the FFSM, C_L for the cLFSM, and $C \setminus C_P$ for the (C\P)FSM.

Definition 3.7. A *fixed fractional stable motion set* (*FFSM set*, in short) of a self-similar mixed moving average X_α given by (3.1) is defined by

$$C_F = \Big\{x \in X : \exists\, c_n = c_n(x) \to 1 \ (c_n \neq 1) : G(x, c_n u) = b_n G(x, u + a_n) + d_n \text{ a.e. } du$$
$$\text{for some } a_n = a_n(c_n, x), b_n = b_n(c_n, x) \neq 0, d_n = d_n(c_n, x) \in \mathbb{R}\Big\}. \quad (3.121)$$

A *cyclic fractional stable motion set* (*cLFSM set*, in short) of a self-similar mixed moving average X_α given by (3.1) is defined by

$$C_L = C_P \setminus C_F, \quad (3.122)$$

where C_P is the PFSM set defined by (3.105) and C_F is the FFSM set defined in (3.121). A *conservative nonperiodic fractional stable motion set* (*(C\P)FSM set*, in short) of a self-similar mixed moving average X_α given by (3.1) is defined by

$$C \setminus C_P, \quad (3.123)$$

where the conservative set C is defined by (3.67).

The PFSM set (3.105) is associated with the PFSM process according to Theorem 3.8, and hence with a periodic flow by Definition 3.5. Both the periodic flow and the function G in (3.105) have a regeneration property, for example, $\psi_c(x) = x$ for $c = c(x) \neq 1$ for the periodic flow. The fixed (identity) flow, on the other hand, is a periodic flow with period 0 for which there are $c_n = c_n(x) \to 1$ $(c_n \neq 1)$ such that $\psi_{c_n}(x) = x$. The definition of a FFSM set in (3.121) similarly requires the regeneration of G at $c_n = c_n(x) \to 1$ $(c_n \neq 1)$.

The next result is analogous to Theorem 3.8.

Theorem 3.9. *A $S\alpha S$ self-similar mixed moving average X_α given by (3.1) with G satisfying (3.26) is:*

- *a FFSM if and only if $C_F = X$ μ-a.e., where C_F is the FFSM set defined in (3.121),*
- *a cLFSM if and only if $C_L = X$ μ-a.e., where C_L is the cLFSM set defined in (3.122),*
- *a (C\P)FSM if and only if $C \setminus C_P = X$ μ-a.e., where $C \setminus C_P$ is the (C\P)FSM set defined in (3.123).*

In the case of a FFSM, the theorem follows from Theorem 10.1 and Corollary 10.1 in Pipiras and Taqqu [36], and Proposition 5.1 in Pipiras and Taqqu [40]. The latter proposition, in fact, states that μ-a.e.,

$$C_F = \Big\{x \in X : G(x,u) = d(u+f)_+^{\kappa} + h(u+f)_-^{\kappa} + g \text{ a.e. } du$$
$$\text{for some reals } d = d(x), f = f(x), g = g(x), h = h(x)\Big\} \quad (3.124)$$

when $\kappa \neq 0$, and

$$C_F = \Big\{x \in X : G(x,u) = d\ln|u+f| + h1_{(0,\infty)}(u+f) + g \text{ a.e. } du$$
$$\text{for some reals } d = d(x), f = f(x), g = g(x), h = h(x)\Big\} \quad (3.125)$$

when $\kappa = 0$. The identification of a FFSM carried out in Pipiras and Taqqu [36] uses these representations of the FFSM set C_F. In the case of a cLFSM, the theorem above is Theorem 5.1 in Pipiras and Taqqu [40]. The case of a (C\P)FSM can be dealt with similarly.

In the next result, we summarize the refined decomposition (3.97) in view of the results of this section.

Theorem 3.10. *Let X_α be a SαS self-similar mixed moving average given by a possibly nonminimal representation (3.1). Suppose that*

$$X_\alpha^D,\ X_\alpha^F,\ X_\alpha^L,\ X_\alpha^{C\setminus P}$$

are the four independent components in the unique decomposition (3.97) of the process X_α obtained by using its minimal representation. Then,

$$X_\alpha^D(t) \stackrel{d}{=} \int_D \int_{\mathbb{R}} G_t(x,u) M(dx,du), \quad (3.126)$$

$$X_\alpha^F(t) \stackrel{d}{=} \int_{C_F} \int_{\mathbb{R}} G_t(x,u) M(dx,du), \quad (3.127)$$

$$X_\alpha^L(t) \stackrel{d}{=} \int_{C_L} \int_{\mathbb{R}} G_t(x,u) M(dx,du), \quad (3.128)$$

$$X_\alpha^{C\setminus P}(t) \stackrel{d}{=} \int_{C\setminus C_P} \int_{\mathbb{R}} G_t(x,u) M(dx,du), \quad (3.129)$$

where $\stackrel{d}{=}$ stands for the equality in the sense of the finite-dimensional distributions and the sets D, C, C_F, C_P, and C_L are defined by (3.67), (3.121), (3.105), and (3.122), respectively.

PROOF: The equality (3.126) follows from Theorem 3.5. The equality (3.127) follows from Corollary 9.1 in Pipiras and Taqqu [36], and will not be proved separately. Consider now the equality (3.128). Let $\widetilde{G}$ be the kernel of a minimal representation (3.108) of the process X_α, and let also $\widetilde{C}_F$, $\widetilde{C}_P$, and $\widetilde{C}_L$ be the sets defined by (3.121), (3.105), and (3.122), respectively, using the kernel function $\widetilde{G}$. Since $C_P = \Phi_1^{-1}(\widetilde{C}_P)$ μ-a.e. by (3.113) and $C_F = \Phi_1^{-1}(\widetilde{C}_F)$ μ-a.e. as shown in the proof of Proposition 7.1 in Pipiras and Taqqu [36], we obtain that

$$C_L = C_P \setminus C_F = \Phi_1^{-1}(\widetilde{C}_P \setminus \widetilde{C}_F) = \Phi_1^{-1}(\widetilde{C}_L) \quad \mu\text{-a.e.}$$

Then, by using (3.109), (3.110) and arguing as in the proof of Theorem 3.5 (see (3.66)), we get that

$$\int_{C_L}\int_{\mathbb{R}} G_t(x,u)M(dx,du) \stackrel{d}{=} \int_{\widetilde{C}_L}\int_{\mathbb{R}} \widetilde{G}_t(\widetilde{x},u)\widetilde{M}(d\widetilde{x},du).$$

Since $\widetilde{G}$ is a kernel of a minimal representation, it is related to a flow in the sense of Definition 3.2. Let $\widetilde{L}$ be the set of the cyclic points of the flow corresponding to the kernel $\widetilde{G}$. By using $\widetilde{P} = \widetilde{C}_P$ in Proposition 3.9 and $\widetilde{F} = \widetilde{C}_F$ in Theorem 10.1 in Pipiras and Taqqu [36], we have $\widetilde{L} = \widetilde{C}_L$ μ-a.e. Then,

$$\int_{C_L}\int_{\mathbb{R}} G_t(x,u)M(dx,du) \stackrel{d}{=} \int_{\widetilde{L}}\int_{\mathbb{R}} \widetilde{G}_t(\widetilde{x},u)\widetilde{M}(d\widetilde{x},du). \tag{3.130}$$

The process on the right-hand side of (3.130) has the distribution of X^L_α by the definition of X^L_α and the uniqueness result in Theorem 3.7.

To show the equality (3.129), observe that by Lemma 3.4 below, we have

$$C_F \subset C_P \subset C.$$

Since $C_P = C_F + C_L$, the sets C_F, C_L and $C \setminus C_P$ are disjoint, and

$$C_F + C_L + C\setminus C_P = C.$$

Hence, the processes on the right-hand side of (3.126)–(3.129) are independent. Since the processes on the left-hand side of (3.126)–(3.129) are also independent, since their sum has the same distribution as the sum of the processes on the right-hand side of (3.126)–(3.129), and since we already showed that the equalities (3.126)–(3.128) hold, we conclude that the equality (3.129) holds as well. □

The following lemma was used in the proof of Theorem 3.10 above.

Lemma 3.4. *We have*

$$C_P \subset C, \tag{3.131}$$

where C_P is the PFSM set (3.105) and C is defined by (3.67).

PROOF: If $x \in C_P$, then by (3.105),

$$G_{rc}(x,rcu) = G(x,rc(1+u)) - G(x,rcu)$$

$$= b(G(x,c(1+u)+a) - G(x,cu+a)) = bG_c(x,cu+a) \quad \text{a.e. } du,$$

for any $c > 0$ and some $r = r(x) \neq 1$, $b = b(x) \neq 0$ and $a = a(x)$. Suppose without loss of generality that $r = r(x) > 1$. Then, by making changes of variables c to rc and u to $u - c^{-1}a$, we obtain that for any $n \in \mathbb{Z}$,

$$\int_{r^n}^{r^{n+1}} dc \int_{\mathbb{R}} du\, c^{-H\alpha}|G_c(x,cu)|^\alpha = r^{1-H\alpha}|b|^\alpha \int_{r^{n-1}}^{r^n} dc \int_{\mathbb{R}} du\, c^{-H\alpha}|G_c(x,cu)|^\alpha$$

and hence

$$\int_{r^n}^{r^{n+1}} dc \int_{\mathbb{R}} du\, c^{-H\alpha} |G_c(x,cu)|^\alpha = r^{(1-H\alpha)n} |b|^{\alpha n} \int_1^r dc \int_{\mathbb{R}} du\, c^{-H\alpha} |G_c(x,cu)|^\alpha.$$

This yields that for $x \in C_P$,

$$\int_0^\infty dc \int_{\mathbb{R}} du\, c^{-H\alpha} |G_c(x,cu)|^\alpha = \sum_{n=-\infty}^{\infty} \int_{r(x)^n}^{r(x)^{n+1}} dc \int_{\mathbb{R}} du\, c^{-H\alpha} |G_c(x,cu)|^\alpha$$

$$= \int_1^{r(x)} dc \int_{\mathbb{R}} c^{-H\alpha} |G_c(x,cu)|^\alpha du \sum_{n=-\infty}^{\infty} r(x)^{(1-H\alpha)n} |b(x)|^{n\alpha} = \infty,$$

since $\sum_{n=-\infty}^{0} r^{(1-H\alpha)n} |b|^{n\alpha} + \sum_{n=1}^{\infty} r^{(1-H\alpha)n} |b|^{n\alpha} = \infty$, which shows that $x \in C$. □

3.2.6 Representations of Processes Related to Conservative Flows

In Theorem 3.6, we provided canonical representations of processes generated by dissipative flows, that is, DFSMs. In this section, we give corresponding representations for FFSMs and PFSMs introduced in Definition 3.5. The next result concerns FFSMs X_α^F. The idea is to randomize the constants a and b in Example 2.5 by replacing them by functions $F_1(x)$ and $F_2(x)$.

Theorem 3.11. *A $S\alpha S$ self-similar mixed moving average X_α is a FFSM if and only if*

$$\begin{cases} \int_X \int_{\mathbb{R}} \Big(F_1(x)((t+u)_+^\kappa - u_+^\kappa) + F_2(x)((t+u)_-^\kappa - u_-^\kappa) \Big) M(dx,du), & \kappa \neq 0, \\ \int_X \int_{\mathbb{R}} \Big(F_1(x) \ln \frac{|t+u|}{|u|} + F_2(x) 1_{(-t,0)}(u) \Big) M(dx,du), & \kappa = 0, \end{cases} \tag{3.132}$$

where $F_1, F_2 : X \to \mathbb{R}$ are some functions and M has the control measure $\nu(dx)du$.

PROOF: By Definition 3.5, a FFSM has a (minimal) representation which is generated by a fixed flow. Denote this representation by $\{G_t\}$ as in Definition 3.2. Then, by using Examples 2.13 and 2.16, the relation (3.25) becomes[10]

$$c^{-\kappa} G(x,cu) = G(x, u + (c^{-1} - 1) g(x)) + j(x) \begin{Bmatrix} c^{-\kappa} - 1, & \kappa \neq 0 \\ \ln c, & \kappa = 0 \end{Bmatrix} \tag{3.133}$$

[10] We use (2.101) and (2.102), but the h is replaced by $\ln c$ since we are considering here a multiplicative flow instead of an additive flow. Note also that $\beta_2 = \kappa$ for the 2-semi-additive functional in Definition 3.2.

for all $c > 0$ a.e. $\mu(dx)du$, which by Lemma 1.1 holds also a.e. $\mu(dx)dudc$. Consider first the case $\kappa \neq 0$. By setting $z = cu$, we obtain that

$$G(x,z) = c^{\kappa}G(x,c^{-1}(z+g(x)) - g(x)) + j(x)(1-c^{\kappa})$$

and hence

$$\begin{aligned} G(x,z) - j(x) &= c^{\kappa}\Big(G(x,c^{-1}(z+g(x)) - g(x)) - j(x)\Big) \\ &= c^{\kappa}F(x,c^{-1}(z+g(x))) \end{aligned}$$

a.e. $\mu(dx)dzdc$ for some function F. By letting

$$\widetilde{G}(x,z) = G(x,z-g(x)) - j(x),$$

we get that $\widetilde{G}(x,z) = c^{\kappa}F(x,c^{-1}z)$ a.e. $\mu(dx)dzdc$. In particular,

$$\widetilde{G}(x,z) = z^{\kappa}(c^{-1}z)^{-\kappa}F(x,c^{-1}z) \quad \text{a.e. } \mu(dx)dzdc$$

when $z > 0$ and hence

$$\widetilde{G}(x,z) = z^{\kappa}v^{-\kappa}F(x,v) \quad \text{a.e. } \mu(dx)dvdz$$

when $z > 0$. By fixing $v = v_0$, for which this equation holds a.e. $\mu(dx)dz$, we get that

$$\widetilde{G}(x,z) = z^{\kappa}F_1(x) \quad \text{a.e. } \mu(dx)dz$$

for some function F_1 when $z > 0$. Similarly, $\widetilde{G}(x,z) = z_-^{\kappa}F_2(x)$ a.e. $\mu(dx)dz$ for some function F_2 when $z < 0$. Hence

$$G(x,z) = F_1(x)(z+g(x))_+^{\kappa} + F_2(x)(z+g(x))_-^{\kappa} + j(x) \tag{3.134}$$

a.e. $\mu(dx)dz$. The result (3.132) of the theorem now follows in the case $\kappa \neq 0$.

When $\kappa = 0$, by making the change of variables $u + c^{-1}g(x) = v$ in (3.133) (and denoting $j(x)$ by $F_1(x)$), we have

$$G(x,cv-g(x)) = G(x,v-g(x)) + F_1(x)\ln c$$

or, by setting $\widetilde{G}(x,z) = G(x,z-g(x))$,

$$\widetilde{G}(x,cv) = \widetilde{G}(x,v) + F_1(x)\ln c$$

a.e. $\mu(dx)dcdv$. Consider now $v > 0$ and write the above relation as

$$\widetilde{G}(x,cv) - F_1(x)\ln cv = \widetilde{G}(x,v) - F_1(x)\ln v$$

a.e. $\mu(dx)dcdv$. By setting

$$\widehat{G}(x,z) = \widetilde{G}(x,z) - F_1(x)\ln z \quad \text{for } x \in X, z > 0,$$

we then have $\widehat{G}(x,cv) = \widehat{G}(x,v)$ a.e. $\mu(dx)dcdv$. By making the change of variables $c = z/v$, we get $\widehat{G}(x,z) = \widehat{G}(x,v)$ a.e. $\mu(dx)dzdv$. Then, by fixing v, we deduce that

$$\widehat{G}(x,z) = F_{2,1}(x)1_{(0,\infty)}(z) \quad \text{a.e. } \mu(dx)dz$$

for some $F_{2,1}$. Going backwards, we get for $z + g(x) > 0$,

$$G(x,z) = \widetilde{G}(x, z+g(x)) = F_{2,1}(x)1_{(0,\infty)}(z+g(x)) + F_1(x)\ln(z+g(x))$$

a.e. $\mu(dx)dz$. When $z + g(x) < 0$, one may deduce similarly that for some function $F_{2,2}$,

$$\begin{aligned} G(x,z) &= F_{2,2}(x)1_{(-\infty,0)}(z+g(x)) + F_1(x)\ln|z+g(x)| \\ &= F_{2,2}(x) - F_{2,2}(x)1_{(0,\infty)}(z+g(x)) + F_1(x)\ln|z+g(x)| \end{aligned}$$

a.e. $\mu(dx)dz$. By combining the previous two relations, we get

$$G(x,z) = (F_{2,1}(x) - F_{2,2}(x))1_{(0,\infty)}(z+g(x)) + F_1(x)\ln|z+g(x)| + F_{2,2}(x) \quad (3.135)$$

a.e. $\mu(dx)dz$. After considering the difference $G(x,t+u) - G(x,u)$, wherein $F_{2,2}(x)$ cancels out, and after making the change of variables $u + g(x)$ to u, this yields the representation (3.132).

Conversely, we have to show that the process given by (3.132) is a FFSM. This follows by observing that $C_F = X$, where C_F is given by (3.124)–(3.125), and Theorem 3.9. □

Example 3.14. Note that when $X = \{1\}$, $\nu(dx) = \delta_{\{1\}}(dx)$ and $\kappa \neq 0$, the process given by (3.132) is the usual LFSM in Example 3.1. When $\kappa = 0$, the process is a linear combination of the stable log-fractional motion in Example 3.7 and the stable Lévy process in Example 2.3. This is as it should be since both are self-similar with

$$\kappa = H - 1/\alpha = 1/\alpha - 1/\alpha = 0.$$

The next result concerns PFSMs. They are obtained as a sum of two independent processes, one with the appearance of a cyclical structure, and the second "fixed" as in Theorem 3.11. We write "appearance" because as stated in Proposition 3.10 below, a FFSM having the representation (3.137) can also be represented by (3.136).

Theorem 3.12. *A $S\alpha S$ self-similar mixed moving average X_α is a PFSM if and only if X_α can be represented by the sum of two independent processes:*

(i) The first process has the representation

$$\int_Z \int_{[0,q(z))} \int_{\mathbb{R}} \Big\{ \Big(b_1(z)^{[v+\ln|t+u|]_{q(z)}} F_1(z, \{v+\ln|t+u|\}_{q(z)}) (t+u)_+^\kappa \\ - b_1(z)^{[v+\ln|u|]_{q(z)}} F_1(z, \{v+\ln|u|\}_{q(z)}) u_+^\kappa \Big)$$

$$+\Big(b_1(z)^{[v+\ln|t+u|]_{q(z)}}F_2(z,\{v+\ln|t+u|\}_{q(z)})\,(t+u)_-^{\kappa}$$

$$-\,b_1(z)^{[v+\ln|u|]_{q(z)}}F_2(z,\{v+\ln|u|\}_{q(z)})\,u_-^{\kappa}\Big)$$

$$+\,1_{\{b_1(z)=1\}}1_{\{\kappa=0\}}F_3(z)\ln\frac{|t+u|}{|u|}\Big\}M(dz,dv,du), \tag{3.136}$$

where $(Z,\mathscr{Z},\sigma)$ *is a standard Lebesgue space,* $b_1(z)\in\{-1,1\}$, $q(z)>0$ *a.e.* $\sigma(dz)$ *and*

$$F_1,F_2:Z\times[0,q(\cdot))\to\mathbb{R},\ F_3:Z\to\mathbb{R}$$

are measurable functions, and M *has the control measure* $\sigma(dz)dvdu$.

(ii) The second process has the representation

$$\begin{cases}\int_Y\int_{\mathbb{R}}\Big(F_1(y)((t+u)_+^{\kappa}-u_+^{\kappa})+F_2(y)((t+u)_-^{\kappa}-u_-^{\kappa})\Big)M(dy,du), & \kappa\neq 0,\\ \int_Y\int_{\mathbb{R}}\Big(F_1(y)\ln\frac{|t+u|}{|u|}+F_2(y)1_{(-t,0)}(u)\Big)M(dy,du), & \kappa=0,\end{cases} \tag{3.137}$$

where $(Y,\mathscr{Y},\nu)$ *is a standard Lebesgue space,* $F_1,F_2:Y\to\mathbb{R}$ *are some functions, and* M *has the control measure* $\nu(dy)du$.

PROOF: Suppose that the process is a PFSM. By Definition 2.3, we have $P=F+L$ and by Definition 3.5, a PFSM is the sum of two independent processes, one of which is a cLFSM and the other is a FFSM. In view of Theorem 3.11, it is enough to show that a cLFSM can be represented as (3.136).

A cLFSM is generated by a cyclic flow $\{\psi_c\}_{c>0}$ on $(X,\mathscr{X},\mu)$. Then, by Proposition 2.7, there are a standard Lebesgue space $(Z,\mathscr{Z},\sigma)$, function $q(z)>0$ and a null-isomorphism

$$\Phi:Z\times[0,q(\cdot))\to X$$

such that

$$\widetilde{\psi}_c(\Phi(z,v))=\Phi(z,\{v+\ln c\}_{q(z)}) \tag{3.138}$$

for all $c>0$ and $(z,v)\in Z\times[0,q(\cdot))$. In other words, the flow $\{\psi_c\}_{c>0}$ on (X,μ) is null-isomorphic to the flow $\{\widetilde{\psi}_c\}_{c>0}$ on $(Z\times[0,q(\cdot)),\sigma(dz)dv)$ defined by

$$\widetilde{\psi}_c(z,v)=(z,\{v+\ln c\}_{q(z)}).$$

(We may suppose that the null sets in Proposition 2.7 are empty because, otherwise, we can replace X by $X\setminus N$ in the definition of the process without changing its distribution.) By replacing x by $\Phi(z,v)$ in (3.25) and using (3.138) which gives $\psi_c\circ\Phi=\Phi\circ\widetilde{\psi}_c$, we get that for all $c>0$,

$$c^{-\kappa}G(\Phi(z,v),cu)=b_c(\Phi(z,v))\left\{\frac{d(\mu\circ\psi_c)}{d\mu}(\Phi(z,v))\right\}^{1/\alpha}\times$$

$$\times\,G\Big(\Phi(\widetilde{\psi}_c(z,v)),u+g_c(\Phi(z,v))\Big)+j_c(\Phi(z,v)) \tag{3.139}$$

a.e. $\sigma(dz)dvdu$. Proceeding as in (3.76), we have

$$\frac{d(\mu\circ\psi_c)}{d\mu}\circ\Phi = \frac{d(\mu\circ\Phi\circ\widetilde{\psi}_c)}{d(\mu\circ\Phi)}$$
$$= \left(\frac{d(\mu\circ\Phi\circ\widetilde{\psi}_c)}{d((\sigma\times\mathbb{L})\circ\widetilde{\psi}_c)}\right)\frac{d((\sigma\times\mathbb{L})\circ\widetilde{\psi}_c)}{d(\sigma\times\mathbb{L})}\frac{d(\sigma\times\mathbb{L})}{d(\mu\circ\Phi)}$$
$$= \left(\frac{d(\mu\circ\Phi)}{d(\sigma\times\mathbb{L})}\circ\widetilde{\psi}_c\right)\frac{d(\sigma\times\mathbb{L})}{d(\mu\circ\Phi)}$$
$$= \left(\frac{d(\mu\circ\Phi)}{d(\sigma\times\mathbb{L})}\circ\widetilde{\psi}_c\right)\left(\frac{d(\mu\circ\Phi)}{d(\sigma\times\mathbb{L})}\right)^{-1},$$

where $\mathbb{L}$ is the Lebesgue measure. Hence, setting

$$\widetilde{G}(z,v,u) = \left\{\frac{d(\mu\circ\Phi)}{d(\sigma\times\mathbb{L})}(z,v)\right\}^{1/\alpha} G(\Phi(z,v),u), \tag{3.140}$$

we obtain that for all $c>0$,

$$c^{-\kappa}\widetilde{G}(z,v,cu) = \widetilde{b}_c(z,v)\widetilde{G}(\widetilde{\psi}_c(z,v),u+\widetilde{g}_c(z,v)) + \widetilde{j}_c(z,v) \tag{3.141}$$

a.e. $\sigma(dz)dvdu$, where

$$\widetilde{b}_c(z,v) = b_c(\Phi(z,v))$$

is a cocycle,

$$\widetilde{g}_c(z,v) = g_c(\Phi(z,v))$$

is a 1-semi-additive functional and

$$\widetilde{j}_c(z,v) = \left\{\frac{d(\mu\circ\Phi)}{d(\sigma\times\mathbb{L})}(z,v)\right\}^{1/\alpha} j_c(\Phi(z,v))$$

is a 2-semi-additive functional for the flow $\widetilde{\psi}_c$. We next consider the case $\kappa\neq 0$ only. The proof in the case $\kappa=0$ can be found in Pipiras and Taqqu [39].

By using (2.93), (2.107), and (2.108) with $\widetilde{\varepsilon}$ replaced by $\widetilde{b}$ and $\widetilde{\varepsilon}_0$ by $\widetilde{b}_1$, and h replaced by $\ln c$ for multiplicative flows, and setting

$$\widehat{G}(z,v,u) = \widetilde{b}(z,v)\Big(\widetilde{G}(z,v,u+\widetilde{g}(z,v)) - \widetilde{j}(z,v)\Big), \tag{3.142}$$

the relation (3.141) can be expressed as, for all $c>0$,

$$\widehat{G}(z,v,cu) = c^{\kappa}\widetilde{b}_1(z)^{[v+\ln c]_{q(z)}}\widehat{G}(z,\{v+\ln c\}_{q(z)},u)$$

a.e. $\sigma(dz)dvdu$. By making the change of variables $cu=w$, we get that for all $c>0$,

$$\widehat{G}(z,v,w) = c^{\kappa}\widetilde{b}_1(z)^{[v+\ln c]_{q(z)}}\widehat{G}(z,\{v+\ln c\}_{q(z)},c^{-1}w)$$

$$= |w^{-1}c|^{\kappa}|w|^{\kappa}\widetilde{b}_1(z)^{[v+\ln|w||w^{-1}c|]_{q(z)}}\widehat{G}(z,\{v+\ln|w||w^{-1}c|\}_{q(z)},c^{-1}w)$$

a.e. $\sigma(dz)dvdw$. By Lemma 1.1, this relation holds a.e. $\sigma(dz)dvdwdc$ as well. Then, for $w>0$, by making the change of variables $c=yw$ and then fixing $y=y_0$, we get

$$\widehat{G}(z,v,w)=w^{\kappa}\widetilde{b}_1(z)^{[v+\ln a_1|w|]_{q(z)}}\widehat{F}_1(z,\{v+\ln a_1|w|\}_{q(z)})$$

a.e. $\sigma(dz)dvdw$, for some $a_1>0$ and function $\widehat{F}_1$. By using identities

$$\{v+\ln a_1|w|\}_{q(z)}=\{\{v+\ln|w|\}_{q(z)}+\ln a_1\}_{q(z)}$$

and

$$[v+\ln a_1|w|]_{q(z)}=[v+\ln|w|]_{q(z)}+[\{v+\ln|w|\}_{q(z)}+\ln a_1]_{q(z)},$$

we can simplify the last relation as

$$\widehat{G}(z,v,w)=w^{\kappa}\widetilde{b}_1(z)^{[v+\ln|w|]_{q(z)}}F_1(z,\{v+\ln|w|\}_{q(z)}) \tag{3.143}$$

a.e. $\sigma(dz)dvdw$, for some function F_1. Similarly, for $w<0$, we may get that

$$\widehat{G}(z,v,w)=w_-^{\kappa}\widetilde{b}_1(z)^{[v+\ln|w|]_{q(z)}}F_2(z,\{v+\ln|w|\}_{q(z)}) \tag{3.144}$$

a.e. $\sigma(dz)dvdw$, for some function F_2. Observe now that by writing characteristic functions and using (3.140) and (3.142),

$$\{X_\alpha(t)\}_{t\in\mathbb{R}}\stackrel{d}{=}\left\{\int_X\int_{\mathbb{R}}(G(x,t+u)-G(x,u))M(dx,du)\right\}_{t\in\mathbb{R}}$$

$$\stackrel{d}{=}\left\{\int_Z\int_{[0,q(\cdot))}\int_{\mathbb{R}}(\widetilde{G}(z,v,t+u)-\widetilde{G}(z,v,u))\widetilde{M}(dz,dv,du)\right\}_{t\in\mathbb{R}}$$

$$\stackrel{d}{=}\left\{\int_Z\int_{[0,q(\cdot))}\int_{\mathbb{R}}(\widehat{G}(z,v,t+u)-\widehat{G}(z,v,u))\widetilde{M}(dz,dv,du)\right\}_{t\in\mathbb{R}}, \tag{3.145}$$

where $\widetilde{M}(dz,dv,du)$ is a $S\alpha S$ random measure on $(Z\times[0,q(\cdot)))\times\mathbb{R}$ with the control measure $\sigma(dz)dvdu$. The result of the theorem when $\kappa\neq 0$ then follows by using (3.143) and (3.144).

The converse of the theorem can be shown by proving that the processes (3.136) and (3.137) are PFSMs. For example, for the process (3.136), note that it is a mixed moving average defined through the kernel function

$$G(z,v,u)=b_1(z)^{[v+\ln|u|]_{q(z)}}\Big(F_1(z,\{v+\ln|u|\}_{q(z)})\,u_+^{\kappa}+F_2(z,\{v+\ln|u|\}_{q(z)})\,u_-^{\kappa}\Big)$$
$$+\ 1_{\{b_1(z)=1\}}1_{\{\kappa=0\}}F_3(z)\ln|u|. \tag{3.146}$$

It satisfies the relation

$$G(z,v,c(z)u)=c(z)^{\kappa}b_1(z)G(z,v,u)+F_3(z)q(z)1_{\{b_1(z)=1\}}1_{\{\kappa=0\}},$$

where $c(z) = e^{q(z)}$. Hence, $C_P = (Z \times [0, q(\cdot)))$ where C_P is the PFSM set defined by (3.105). Hence, the process (3.136) is a PFSM by using Theorem 3.8. □

The representation (3.136) is not specific to processes generated by cyclic flows. The next result shows that FFSMs (they are generated by fixed flows) can also be represented by (3.136).

Proposition 3.10. *A FFSM having the representation (3.137) can be represented by (3.136).*

PROOF: Consider first the case $\kappa \neq 0$. Taking $b_1(z) \equiv 1$, $q(z) \equiv 1$, $F_1(z,v) \equiv F_1(z)$, $F_2(z,v) \equiv F_2(z)$, and $F_3(z) \equiv 0$ in (3.136), we obtain the process

$$\int_Z \int_0^1 \int_{\mathbb{R}} 1_{[0,1)}(v) \Big(F_1(z)((t+u)_+^\kappa - u_+^\kappa) + F_2(z)((t+u)_-^\kappa - u_-^\kappa) \Big) M(dz, dv, du).$$

Since the kernel above involves the variable v only through the indicator function $1_{[0,1)}(v)$ and since the control measure of $M(dz, dv, du)$ in variable v is dv, the latter process has the same finite-dimensional distributions as

$$\int_Z \int_{\mathbb{R}} \Big(F_1(z)((t+u)_+^\kappa - u_+^\kappa) + F_2(z)(t+u)_-^\kappa - u_-^\kappa) \Big) M(dz, du),$$

which is the representation (3.137) of a FFSM when $\kappa \neq 0$. In the case $\kappa = 0$, one can arrive at the same conclusion by taking

$$b_1(z) \equiv 1,\ q(z) \equiv 1,\ F_1(z,v) = F_2(z),\ F_2(z,v) = 0 \text{ and } F_3(z) = F_1(z). \qquad \square$$

The next corollary which is an immediate consequence of Theorem 3.12 and Proposition 3.10 states, as indicated earlier, that it is not necessary to include Part (*ii*) in Theorem 3.12.

Corollary 3.5. *A $S\alpha S$ self-similar mixed moving average X_α is a PFSM if and only if it can be represented by (3.136).*

The next auxiliary result is useful in showing that a PFSM is, in fact, a cLFSM.

Lemma 3.5. *If the process X_α has the representation (3.136) and $C_F = \emptyset$ a.e. where the set C_F is defined by (3.121) using the representation (3.136), then X_α is a cLFSM. Moreover, to show that $C_F = \emptyset$ a.e., it is enough to prove that $(z,0) \notin C_F$ a.e., that is, $(z,v) \notin C_F$ a.e. when $v = 0$.*

PROOF: If X_α is represented by (3.136), then by Corollary 3.5, X_α is a PFSM. By Definition 3.5, the PFSM set C_P associated with the representation (3.136) is the whole space a.e. Since $C_L = C_P \setminus C_F$, the assumption $C_F = \emptyset$ a.e. implies that C_L is the whole space a.e. as well. Therefore, X_α is a cLFSM by Definition 3.5.

The last statement of the lemma follows if we show that $(z,v) \in C_F$ if and only if $(z,0) \in C_F$. The representation (3.136) of the process X_α is defined through the kernel function in (3.146). Observe that

$$G(z,v,u) = e^{-\kappa v} G(z,0,e^v u) - 1_{\{b_1(z)=1\}} 1_{\{\kappa=0\}} F_3(z) v \tag{3.147}$$

for all z,v,u, and that by making the change of variables $e^v u = w$, $v = v$, and $z = z$ in (3.147),

$$G(z,0,w) = e^{\kappa v} G(z,v,e^{-v} w) + 1_{\{b_1(z)=1\}} 1_{\{\kappa=0\}} F_3(z) v e^{\kappa v}$$

for all z,v,w. By using these relations and the definition (3.121) of C_F, we obtain that there is $c_n(z,v) \to 1$ ($c_n(z,v) \neq 1$) such that

$$G(z,v,c_n(z,v)u) = b_n(z,v) G(z,v,u + a_n(z,v)) + d_n(z,v), \quad \text{a.e. } du,$$

for some $a_n(z,v), b_n(z,v) \neq 0, d_n(z,v)$, if and only if there is $\widetilde{c}_n(z) \to 1$ ($\widetilde{c}_n(z) \neq 1$) such that

$$G(z,0,\widetilde{c}_n(z)u) = \widetilde{b}_n(z) G(z,0,u + \widetilde{a}_n(z)) + \widetilde{d}_n(z), \quad \text{a.e. } du,$$

for some $\widetilde{a}_n(z), \widetilde{b}_n(z) \neq 0, \widetilde{d}_n(z)$. This shows that $(z,v) \in C_F$ if and only if $(z,0) \in C_F$.
□

Further study of PFSMs and cLFSMs can be found in Pipiras and Taqqu [39], including the study of integrand spaces, uniqueness questions, and examples. We include some of those results and examples next. In particular, there are many other ways to represent processes (3.136) as indicated in the following lemma.

Lemma 3.6. *Other representations for the process X_α in (3.136) are as follows:*

$$X_\alpha(t) \stackrel{d}{=} \int_Z \int_0^{q(z)} \int_{\mathbb{R}} e^{-\kappa v} (K(z, e^v(t+u)) - K(z, e^v u)) M(dz, dv, du) \tag{3.148}$$

$$\stackrel{d}{=} \int_Z \int_1^{e^{q(z)}} \int_{\mathbb{R}} w^{-H} (K(z, w(t+u)) - K(z, wu)) M(dz, dw, du) \tag{3.149}$$

$$\stackrel{d}{=} \int_Z \int_1^{e^{-q(z)}} \int_{\mathbb{R}} w^{H-\frac{2}{\alpha}} (K(z, w^{-1}(t+u)) - K(z, w^{-1}u)) M(dz, dw, du) \tag{3.150}$$

$$\stackrel{d}{=} \int_Z \int_1^{e^{q(z)}} \int_{\mathbb{R}} w^{-H-\frac{1}{\alpha}} (K(z, wt+u)) - K(z,u)) M(dz, dw, du) \tag{3.151}$$

$$\stackrel{d}{=} \int_Z \int_1^{e^{q(z)}} \int_{\mathbb{R}} w^{H-\frac{1}{\alpha}} (K(z, w^{-1}t+u)) - K(z,u)) M(dz, dw, du), \tag{3.152}$$

where $\stackrel{d}{=}$ denotes the equality in the sense of the finite-dimensional distributions, $M(dz,dv,du)$ has the control measure $\sigma(dz)dvdu$, and

$$K(z,u) = b_1(z)^{[\ln|u|]_{q(z)}} \Big(F_1(z, \{\ln|u|\}_{q(z)}) u_+^\kappa + F_2(z, \{\ln|u|\}_{q(z)}) u_-^\kappa \Big)$$
$$+ 1_{\{b_1(z)=1\}} 1_{\{\kappa=0\}} F_3(z) \ln|u|. \tag{3.153}$$

PROOF: The lemma follows by making proper changes of variables in (3.136). □

It can be shown that a self-similar mixed moving average given by the representation (3.136) is well defined if and only if

$$\int_Z \int_{\mathbb{R}} \int_1^{e^{q(z)}} h^{-\alpha H-1} |K(z,u+h) - K(u)|^\alpha \sigma(dz) du dh < \infty, \tag{3.154}$$

where K is defined by (3.153) (see Pipiras and Taqqu [39], Proposition 5.2). Sufficient conditions for (3.154) to hold in the special case $Z = \{1\}$, $\sigma(dz) = \delta_{\{1\}}(dz)$, are provided in the next result.

Lemma 3.7. *Let $\alpha \in (0,2)$, $H \in (0,1)$ and $\kappa = H - 1/\alpha$, and let K be defined by (3.153) with $Z = \{1\}$ and $\sigma(dz) = \delta_{\{1\}}(dz)$.*

(i) Suppose that $\kappa < 0$. If $F_1, F_2 : [0,q) \to \mathbb{R}$ are such that F_1, F_2 are absolutely continuous with derivatives F_1', F_2', and

$$\sup_{u \in [0,q)} |F_i(u)| \le C, \quad i = 1,2, \tag{3.155}$$

$$\operatorname{ess} \sup_{u \in [0,q)} |F_i'(u)| \le C, \quad i = 1,2, \tag{3.156}$$

then (3.154) holds.

(ii) Suppose that $\kappa \ge 0$. If, in addition to (i),

$$F_i(0) = b_1 F_i(q-), \quad i = 1,2, \tag{3.157}$$

then (3.154) holds.

PROOF: See the proof of Lemma 5.1 in Pipiras and Taqqu [39]. □

The next example provides functions F_1, F_2 satisfying the conditions in Lemma 3.7 above, and hence a well-defined PFSM. We also show that the constructed PFSM is, in fact, a cLFSM.

Example 3.15. Let $\alpha \in (0,2)$ and $H \in (0,1)$ and hence $\kappa = H - 1/\alpha \in (-1/\alpha, 1 - 1/\alpha)$. The process

$$X_\alpha(t) = \int_0^1 \int_{\mathbb{R}} \left(F(\{v + \ln|t+u|\}_1)(t+u)_+^\kappa - F(\{v + \ln|u|\}_1) u_+^\kappa \right) M(dv, du) \tag{3.158}$$

has the representation (3.136) with $Z = \{1\}$, $\sigma(dz) = \delta_{\{1\}}(dz)$, $b_1(1) = 1$, $q(1) = 1$, $F_1(1,u) = F(u)$, $F_2(1,u) = 0$ and $F_3(1) = 0$. It is well defined by Lemma 3.7 if the function $F : [0,1) \to \mathbb{R}$ satisfies the conditions (3.155) and (3.156) when $\kappa < 0$ and, in addition, the condition (3.157) when $\kappa \ge 0$. We can take, for example,

$$F(u) = u, \quad u \in [0,1), \tag{3.159}$$

when $\kappa < 0$, and

$$F(u) = u 1_{[0,1/2)}(u) + (1-u) 1_{[1/2,1)}(u), \quad u \in [0,1), \tag{3.160}$$

when no additional conditions on κ are imposed. (The function F satisfies (3.157) because $F(0) = 0 = F(1-)$.)

We now show that the PFSM (3.158) with (3.159) or (3.160) above is, in fact, a cLFSM. The PFSM (3.158) is defined through the kernel function

$$G(v,u) = F(\{v + \ln|u|\}_1) u_+^{\kappa},$$

where F is given by (3.159) or (3.160). By Lemma 3.5, it is enough to show that $0 \notin C_F$ where the set C_F is defined by (3.121). If $0 \in C_F$, then there is $c_n \to 1$ ($c_n \neq 1$) such that

$$F(\{\ln|c_n u|\}_1) u_+^{\kappa} = b_n F(\{\ln|u + a_n|\}_1)(u + a_n)_+^{\kappa} + d_n \quad \text{a.e. } du, \tag{3.161}$$

for some $a_n, b_n \neq 0, d_n$. By taking large enough negative u such that $u_+^{\kappa} = 0$ and $(u + a_n)_+^{\kappa} = 0$, we get that $d_n = 0$ and hence (3.161) becomes

$$F(\{\ln|c_n u|\}_1) u_+^{\kappa} = b_n F(\{\ln|u + a_n|\}_1)(u + a_n)_+^{\kappa} \quad \text{a.e. } du. \tag{3.162}$$

We shall distinguish between the cases $\kappa < 0$ and $\kappa \geq 0$. Observe that we need to consider the functions (3.159) and (3.160) when $\kappa < 0$, and only the function (3.160) when $\kappa \geq 0$.

If $\kappa < 0$, since F is bounded and not identically zero, by letting $u \to 0$ in (3.162), we obtain that $a_n = 0$. Hence, when $\kappa < 0$, $F(\{\ln|c_n u|\}_1) 1_{(0,\infty)}(u) = b_n F(\{\ln|u|\}_1) 1_{(0,\infty)}(u)$ a.e. du or, by setting $e_n = \ln c_n$ and $v = \ln u$,

$$F(\{e_n + v\}_1) = b_n F(\{v\}_1) \quad \text{a.e. } dv, \tag{3.163}$$

for $e_n \to 0$ ($e_n \neq 0$) and $b_n \neq 0$. Neither of the functions F in (3.159) or (3.160) satisfies the relation (3.163). For example, if the function (3.159) satisfies (3.163), then $e_n + v = b_n v$ for $v \in [0, 1 - e_n)$ and some $0 < e_n < 1$ which is a contradiction (for example, if $v = 0$, we get $e_n = 0$). If $\kappa \geq 0$, we can also get $a_n = 0$ in (3.162). If, for example, $a_n < 0$, then since $(u + a_n)_+^{\kappa} = 0$ for $u \in [0, -a_n)$, we have $F(\{\ln|c_n u|\}_1) = 0$ for $u \in [0, -a_n)$ or that the function $F(v) = 0$ on an interval of $[0,1)$. The function (3.160) does not have this property.

3.2.7 Summary

The following theorem is essentially a reformulation of Theorem 3.10 and some other previous results. It provides a decomposition of self-similar mixed moving averages which is more refined than that based on Hopf decomposition in Section 3.2.2. As in Section 3.2.5, we will say that a decomposition of a process X_α

obtained from its representation (3.1) is unique in distribution if the distribution of its components does not depend on the representation (3.1). We will also say that a process does not have a PFSM component if it cannot be expressed as the sum of two independent processes where one process is a PFSM.

Theorem 3.13. *Let X_α be a $S\alpha S$ self-similar mixed moving average given by a possibly nonminimal representation (3.1). Then, the process X_α can be decomposed uniquely in distribution into four independent processes*

$$X_\alpha \stackrel{d}{=} X_\alpha^D + X_\alpha^F + X_\alpha^L + X_\alpha^{C\setminus P}, \tag{3.164}$$

where

$$X_\alpha^D(t) = \int_D \int_{\mathbb{R}} G_t(x,u) M(dx,du), \tag{3.165}$$

$$X_\alpha^F(t) = \int_{C_F} \int_{\mathbb{R}} G_t(x,u) M(dx,du), \tag{3.166}$$

$$X_\alpha^L(t) = \int_{C_L} \int_{\mathbb{R}} G_t(x,u) M(dx,du), \tag{3.167}$$

$$X_\alpha^{C\setminus P}(t) = \int_{C\setminus C_P} \int_{\mathbb{R}} G_t(x,u) M(dx,du). \tag{3.168}$$

The sets D and C are defined by (3.67), namely,

$$D = \{x \in X : I(x) < \infty\}, \quad C = \{x \in X : I(x) = \infty\}, \tag{3.169}$$

where

$$I(x) = \int_0^\infty c^{-H\alpha} \int_{\mathbb{R}} |G(x,c(1+u)) - G(x,cu)|^\alpha \, du \, dc.$$

The sets C_F, C_P, and C_L are defined by (3.121), (3.105), and (3.122), namely,

$$C_F = \Big\{ x \in X : \exists\, c_n = c_n(x) \to 1 \ (c_n \neq 1) : G(x,c_n u) = b_n G(x,u+a_n) + d_n \ \text{a.e. } du$$
$$\text{for some } a_n = a_n(c_n,x), b_n = b_n(c_n,x) \neq 0, d_n = d_n(c_n,x) \in \mathbb{R} \Big\}, \tag{3.170}$$

$$C_P = \Big\{ x \in X : \exists\, c = c(x) \neq 1 : G(x,cu) = b G(x,u+a) + d \ \text{ a.e. } du$$
$$\text{for some } a = a(c,x), b = b(c,x) \neq 0, d = d(c,x) \in \mathbb{R} \Big\}, \tag{3.171}$$

and

$$C_L = C_P \setminus C_F.$$

Moreover:

(i) The process X_α^D is generated by a dissipative flow (that is, a DFSM) and has the canonical representation given in Theorem 3.6, namely,

$$\{X_\alpha(t)\}_{t\in\mathbb{R}} \stackrel{d}{=} \left\{ \int_Y \int_{\mathbb{R}} \int_{\mathbb{R}} e^{-\kappa s}(F(y,e^s(t+u)) - F(y,e^s u))M(dy,ds,du) \right\}_{t\in\mathbb{R}}, \tag{3.172}$$

where $F: Y\times\mathbb{R}\to\mathbb{R}$ is some function and M is a SαS random measure on $Y\times\mathbb{R}\times\mathbb{R}$ with the control measure $m(dy,ds,du) = \nu(dy)dsdu$.

(ii) The process X_α^F is a FFSM and has the representation (3.132), namely,

$$\begin{cases} \int_X \int_{\mathbb{R}} \Big(F_1(x)((t+u)_+^\kappa - u_+^\kappa) + F_2(x)((t+u)_-^\kappa - u_-^\kappa)\Big)M(dx,du), & \kappa\neq 0,\\ \int_X \int_{\mathbb{R}} \Big(F_1(x)\ln\frac{|t+u|}{|u|} + F_2(x)1_{(-t,0)}(u)\Big)M(dx,du), & \kappa = 0, \end{cases} \tag{3.173}$$

where $F_1, F_2: X\to\mathbb{R}$ are some functions and M has the control measure $\nu(dx)du$.

(iii) The process X_α^L is a cLFSM, and the sum $X_\alpha^P = X_\alpha^F + X_\alpha^L$ is a PFSM. Both processes have a representation (3.136).

(iv) The process $X_\alpha^{C\setminus P}$ is a self-similar mixed moving average without a PFSM component, that is, a $(C\setminus P)$FSM.

If the process X_α is generated by a flow $\{\psi_c\}_{c>0}$, then the sets D and C are identical (a.e.) to the dissipative and the conservative parts of the flow $\{\psi_c\}_{c>0}$.

If, in addition, the representation of the process X_α is minimal, then the sets C_P, C_F and C_L are the sets of the periodic, fixed, and cyclic points of the flow $\{\psi_c\}_{c>0}$, respectively.

Remark 3.10. Although $C\setminus C_P$ is clearly disjoint from C_P, the process $X_\alpha^{C\setminus P}$ is defined through the integral representation (3.168). It is therefore not obvious a priori that the process $X_\alpha^{C\setminus P}$ cannot be represented as the sum of two independent processes, one of which is a PFSM, namely a process with a representation $\int_{\widetilde{C}_P} \widetilde{G}_t(\widetilde{x},u)\widetilde{M}(d\widetilde{x},du)$ involving a possibly different kernel function $\widetilde{G}$.

PROOF: The uniqueness of the decomposition (3.164) into four independent component follows by using Theorem 3.10 and the uniqueness result in Theorem 3.7. Parts (i) follows from Theorem 3.5. Part (ii) follow from Theorem 9.1 in Pipiras and Taqqu [36]. Part (iii) is a consequence of the equalities (3.127) and (3.128) in Theorem 3.10 and Definition 3.5.

To show that the process $X_\alpha^{C\setminus P}$ does not have a PFSM component, we argue by contradiction. Suppose on the contrary that $X_\alpha^{C\setminus P}$ has a PFSM component, that is,

$$X_\alpha^{C\setminus P} \stackrel{d}{=} V + W,$$

where V and W are independent, and W is a PFSM. Let

$$G^{C\setminus P}: (C\setminus P)\times\mathbb{R}\to\mathbb{R} \quad \text{and} \quad F: Y\times\mathbb{R}\to\mathbb{R}$$

be the kernel functions in the representation of $X_\alpha^{C\setminus P}$ and W, respectively, where the integral representation of W is equipped with the control measure $\sigma(dy)du$. We now apply Proposition 3.3, which concerns processes expressed through two different representations, in our case one with kernel $G^{C\setminus P}$ and the other with kernel F. By using that proposition, there are functions

$$\Phi_1 : Y \to C \setminus C_P, \quad h : Y \to \mathbb{R} \setminus \{0\} \quad \text{and} \quad \Phi_2, \Phi_3 : Y \to \mathbb{R}$$

such that

$$F(y,u) = h(y) G^{C\setminus P}(\Phi_1(y), u + \Phi_2(y)) + \Phi_3(y) \quad \text{a.e. } \sigma(dy)du \tag{3.174}$$

and therefore

$$G^{C\setminus P}(\Phi_1(y), u) = (h(y))^{-1} F(y, u - \Phi_2(y)) - (h(y))^{-1} \Phi_3(y) \quad \text{a.e. } \sigma(dy)du. \tag{3.175}$$

Since F is the kernel function of a PFSM, it satisfies by (3.117),

$$F(y, c(y)u) = b(y) F(y, u + a(y)) + d(y) \quad \text{a.e. } \sigma(dy)du, \tag{3.176}$$

for some

$$c(y) > 0 \ (c(y) \neq 1), \ b(y) \neq 0, \ a(y), d(y) \in \mathbb{R}.$$

Then, by replacing u by $c(y)u$ in (3.175) and by using (3.176) and (3.174), we get that

$$G^{C\setminus P}(\Phi_1(y), c(y)u) = B(y) G^{C\setminus P}(\Phi_1(y), u + A(y)) + D(y) \quad \text{a.e. } \sigma(dy)du, \tag{3.177}$$

for some $B(y) \neq 0$, $A(y), D(y) \in \mathbb{R}$. Since $\sigma(dy)$ is not a zero measure, relation (3.177) contradicts the fact that $\Phi_1(y) \in C \setminus C_P$ in view of the definitions of the set C_P in (3.105).

The last two statements of the theorem follow from the proof of Theorem 5.3 in Pipiras and Taqqu [35], Theorem 10.1 in Pipiras and Taqqu [36] and Propositions 3.9. □

3.2.8 An Example of a Conservative Nonperiodic Fractional Stable Motion

An example of a conservative nonperiodic fractional stable motion $X_\alpha^{C\setminus P}$, that is, a (C\P)FSM or the "fourth type process" in (3.164), can be constructed as follows (see also Pipiras and Taqqu [40], Section 7). Let $\{W(t)\}_{t\in\mathbb{R}}$ be a stationary process which has càdlàg (that is, right continuous and with limits from the left) paths, satisfies $E|W(t)|^\alpha < \infty$,

$$E|W(t) - W(s)|^\alpha \leq C|t-s|^p, \quad s,t \in \mathbb{R}, \tag{3.178}$$

for some $\alpha \in (0,2)$, $p > 0$, $P(|W(t)| < c) < 1$ for all $c > 0$ and is ergodic. Let

$$\Omega = \{w : w(t),\ t \in \mathbb{R},\ \text{is càdlàg}\}$$

be the space of càdlàg functions on $\mathbb{R}$. Let $P(dw)$ be the probability measure on Ω corresponding to the process W. Consider the process

$$X_\alpha(t) = \int_\Omega \int_\mathbb{R} \Big(G(w,t+u) - G(w,u)\Big) M(dw,du)$$

as in (3.1) but where the space X is the sample space Ω endowed with the probability measure P. This measure enters in the specification of the control measure $m(dw,du)$ of $M(dw,du)$ by setting

$$m(dw,du) = P(dw)du.$$

We also set

$$G(w,t) = |w|^\kappa F(w, \ln|t+u|), \tag{3.179}$$

where $\kappa = H - 1/\alpha$ as in (3.24) and $F(w,v) = w(v)$, that is, the value of the function w at time v. Hence

$$G(w,t) = |w|^\kappa w(\ln|t+u|).$$

Thus, the process is

$$\begin{aligned} X_\alpha(t) &= \int_\Omega \int_\mathbb{R} \Big(|t+u|^\kappa F(w,\ln|t+u|) - |u|^\kappa F(w,\ln|u|)\Big) M(dw,du), \\ &= \int_\Omega \int_\mathbb{R} \Big(|t+u|^\kappa w(\ln|t+u|) - |u|^\kappa w(\ln|u|)\Big) M(dw,du), \quad t \in \mathbb{R}, \end{aligned} \tag{3.180}$$

where the control measure of $M(dw,du)$ is $P(dw)du$.

Lemma 3.8. *The process X_α in (3.180) is well defined for*

$$H \in (0, \min\{p,1\}) \quad \textit{and} \quad \alpha \in (0,2)$$

under the assumption (3.178). Moreover, the SαS process X_α is a H-self-similar mixed moving average.

PROOF: The result follows since, by using (3.178) and stationarity of W,

$$\begin{aligned} &\int_\Omega \int_\mathbb{R} \Big| |t+u|^\kappa F(w,\ln|t+u|) - |u|^\kappa F(w,\ln|u|) \Big|^\alpha P(dw)du \\ &= \int_\mathbb{R} E\Big| |t+u|^\kappa W(\ln|t+u|) - |u|^\kappa W(\ln|u|)\Big|^\alpha du \\ &\le 2^\alpha \int_\mathbb{R} |t+u|^{\kappa\alpha} E\Big| W(\ln|t+u|) - W(\ln|u|)\Big|^\alpha du \\ &\quad + 2^\alpha \int_\mathbb{R} E|W(\ln|u|)|^\alpha \Big| |t+u|^\kappa - |u|^\kappa\Big|^\alpha du \end{aligned}$$

$$\leq 2^\alpha C \int_{\mathbb{R}} |t+u|^{\kappa\alpha} \Big| \ln|t+u| - \ln|u| \Big|^{p\alpha} du + 2^\alpha C \int_{\mathbb{R}} \Big| |t+u|^\kappa - |u|^\kappa \Big|^\alpha du < \infty,$$

when $\kappa\alpha - p\alpha + 1 = (H - 1/\alpha)\alpha - p\alpha + 1 = \alpha(H-p) < 0$ and $H < 1$.

To see the H-self-similarity, note that for $c > 0$,

$$\int_\Omega \int_{\mathbb{R}} \Big(|ct+u|^\kappa F(w, \ln|ct+u|) - |u|^\kappa F(w, \ln|u|) \Big) M(dw, du)$$

$$= \int_\Omega \int_{\mathbb{R}} c^\kappa \Big(|t+v|^\kappa F(w, \ln c + \ln|t+v|) - |v|^\kappa F(w, \ln c + \ln|v|) \Big) M(dw, d(cv))$$

$$\stackrel{d}{=} \int_\Omega \int_{\mathbb{R}} c^\kappa \Big(|t+v|^\kappa F(w, \ln c + \ln|t+v|) - |v|^\kappa F(w, \ln c + \ln|v|) \Big) c^{1/\alpha} M(dw, dv)$$

$$\stackrel{d}{=} \int_\Omega \int_{\mathbb{R}} c^H \Big(|t+v|^\kappa F(w, \ln|t+v|) - |v|^\kappa F(w, \ln|v|) \Big) M(dw, dv),$$

where we made the change of variables $u = cv$ and used the stationarity of the process $F(w,s) = w(s)$. □

The fact that X_α is a (C\P)FSM, that is, the decomposition (3.164) of X_α contains only the fourth term $X_\alpha^{C\setminus P}$, is proved as follows. First, the process X_α is generated by a conservative flow. Indeed, in view of (3.179), we have

$$c^{-\kappa} G(w, cu) = G(\psi_c(w), u), \quad c > 0,$$

where

$$\psi_c : w(v), v \in \mathbb{R} \mapsto w(v + \ln c), v \in \mathbb{R}$$

is a measurable flow on Ω. Since the process $W(t)$, $t \in \mathbb{R}$, is stationary, the flow $\{\psi_c\}_{c>0}$ is measure preserving. Then, in the decomposition $D = \cup_{k=-\infty}^{\infty} (\phi_c)^k(B)$ of the dissipative part of the multiplicative flow $\{\psi_c\}_{c>0}$ with a wandering set B (see Section 2.3.3), each set $(\phi_c)^k(B)$ has the same measure. Since the measure P on Ω is finite, there can be no wandering set B of positive measure and hence the flow $\{\psi_c\}_{c>0}$ is conservative. Then, X_α is a (C\P)FSM as long as $C_P = \emptyset$ a.e. $P(dw)$, where C_P is defined in (3.105), that is, in view of (3.179),

$$C_P = \Big\{ w \in \Omega : \exists c \neq 1, a, b \neq 0, d : |cu|^\kappa w(\ln|cu|) = b|u+a|^\kappa w(\ln|u+a|), \quad \forall u \Big\}, \tag{3.181}$$

where the "a.e. du" condition is replaced by the "$\forall u$" condition since the functions w are càdlàg. Informally, the measure of the set C_P is zero since the functions w in C_P are "regenerative" (that is, $w(\ln|cu|) = w(\ln c + \ln|u|)$ is related back to $w(\ln|u|)$), which can be shown to be inconsistent with the assumptions on the process W (or its measure P). This approach is outlined in the proof of the next result.

Lemma 3.9. *If C_P is the PFSM set (3.181) associated with the representation (3.180) of the process X_α, then $C_P = \emptyset$ a.e. $P(dw)$.*

PROOF: We consider only the case $\kappa > 0$. For the case $\kappa \leq 0$, see Pipiras and Taqqu [40], Section 7.

We first examine the case when $b \neq 1$ in (3.181). By using (3.107), we can express the equation in (3.181) as

$$|cu+g|^{\kappa}w(\ln|cu+g|)+f = b\Big(|u+g|^{\kappa}w(\ln|u+g|)+f\Big), \tag{3.182}$$

for some $c > 1, b \neq 0, f, g \in \mathbb{R}$. Setting

$$\widetilde{w}(v) = e^{-\kappa v}\Big(|e^{v}+g|^{\kappa}w(\ln|e^{v}+g|)+f\Big) \tag{3.183}$$

and $\widetilde{c} = \ln c > 0$, we have from (3.182) with $u = e^{v}$ that

$$\widetilde{w}(v+\widetilde{c}) = \widetilde{b}\widetilde{w}(v), \quad v \in \mathbb{R}, \tag{3.184}$$

where $\widetilde{b} = bc^{\kappa}$. Observe also that by making the change of variables $v = \ln(e^{u}-g)$ in (3.183) for large v, we have

$$w(u) = e^{-\kappa u}\Big((e^{u}-g)^{\kappa}\widetilde{w}(\ln(e^{u}-g))-f\Big) \tag{3.185}$$

for large u. We next consider two cases separately, $|\widetilde{b}| \leq 1$ and $|\widetilde{b}| > 1$, and prove that the P-probability of w satisfying (3.182) is zero.

If $|\widetilde{b}| \leq 1$ in (3.184), then $|\widetilde{w}(v)|$ is bounded for large v. Indeed, if $|\widetilde{b}| = 1$, then $|\widetilde{w}(v)|$ is periodic with period $\widetilde{c}$ and, being càdlàg, it has to be bounded. If $|\widetilde{b}| < 1$, then $|\widetilde{w}(v)| \to 0$ as $v \to \infty$ because $|\widetilde{w}(v+n\widetilde{c})| = |\widetilde{b}|^{n}|\widetilde{w}(v)|$ and $|\widetilde{b}|^{n} \to 0$ as $n \to \infty$. By (3.185), since $\kappa > 0$, we obtain that $|w(u)|$ is bounded for large u as well. It is then enough to show that $P(A) = 0$ where

$$A = \{w : |w(u)| \text{ is bounded for large } u\}.$$

Observe that $P(A) \leq \sum_{n=1}^{\infty} P(B_n)$, where $B_n = \{w : |w(u)| < n \text{ for large } u\}$. When $w \in B_n$, we have

$$\frac{1}{T}\int_0^T 1_{\{|w(u)|<n\}}du \to 1,$$

as $T \to \infty$. But by the assumptions on W, namely ergodicity and $P(|w(0)| < c) < 1$ for any $c > 0$, we have

$$\frac{1}{T}\int_0^T 1_{\{|w(u)|<n\}}du \to P(|w(0)| < n) < 1 \quad \text{a.e. } P(dw). \tag{3.186}$$

Since, on the other hand, $\frac{1}{T}\int_0^T 1_{\{|w(u)|<n\}}du \to 1$ on B_n, the relation (3.186) implies that $P(B_n) = 0$, and hence $P(A) = 0$.

Suppose now that $|\widetilde{b}| > 1$ in (3.184). We have either (i) $\widetilde{w}(v) = 0$ for $v \in [0, \widetilde{c}]$, or (ii) $\inf\{|\widetilde{w}(v)| : v \in A\} > 0$ for $A \subset [0, \widetilde{c}]$ of positive Lebesgue measure. In the case (i), (3.184) implies that $\widetilde{w}(v) = 0$ for all v and hence, by (3.185), $w(u) = -fe^{-\kappa u}$ for large u. That is, $|w(u)|$ is bounded for large u, and we deduce as above that the P-probability of such w's is zero. Consider now the case (ii). Since $|\widetilde{b}| > 1$, we get that

$$\inf\{|\widetilde{w}(v)| : v \in A + n\widetilde{c}\} \to \infty, \quad \text{as } n \to \infty.$$

Using (3.183), since $\kappa > 0$ (and hence $fe^{-\kappa u} \to 0$ as $u \to \infty$), this yields that

$$\inf\{|w(\ln(e^{v} + g))| : v \in A + n\widetilde{c}\} \to \infty, \quad \text{as } n \to \infty. \tag{3.187}$$

We shall argue that the P-probability of such w's is zero as well. Indeed, note that the relation (3.187) implies, for fixed u_0,

$$\frac{1}{T}\int_{u_0}^{T} |w(\ln(e^{u} + g))|^{\alpha} du \to \infty,$$

as $T \to \infty$. Making the change of variables $\ln(e^{u} + g) = v$, we also obtain that

$$\frac{1}{T}\int_{\ln(e^{u_0}+g)}^{\ln(e^{T}+g)} |w(v)|^{\alpha} \frac{e^{v}}{e^{v} - g} dv \to \infty$$

and hence that

$$\frac{1}{T}\int_{0}^{T} |w(u)|^{\alpha} du \to \infty.$$

However, by ergodicity and the assumption $E|W(0)|^{\alpha} < \infty$, we have

$$\frac{1}{T}\int_{0}^{T} |w(u)|^{\alpha} du \to E|W(0)|^{\alpha} < \infty \quad \text{a.e. } P(dw).$$

We thus conclude as above that the P-probability of w's satisfying (3.187) is zero.

If $b = 1$ in (3.181), by using (3.106), we get

$$|cu + g|^{\kappa} w(\ln|cu + g|) = |u + g|^{\kappa} w(\ln|u + g|) + d, \tag{3.188}$$

for some $c > 1, g, d \in \mathbb{R}$. Setting

$$\widetilde{w}(v) = |e^{v} + g|^{\kappa} w(\ln|e^{v} + g|) \tag{3.189}$$

and $\widetilde{c} = \ln c > 0$, we deduce from (3.188) with $u = e^{v}$ that

$$\widetilde{w}(v + \widetilde{c}) = \widetilde{w}(v) + d, \quad v \in \mathbb{R}. \tag{3.190}$$

The function $\widetilde{w}$ is bounded on $[0, \widetilde{c}]$ since it is càdlàg and in view of (3.190), we get

$$|\widetilde{w}(v)| \le C|v|, \tag{3.191}$$

for large v and some constant $C = C(w) > 0$. Substituting (3.189) into (3.191), and since $\kappa > 0$, we get that $w(v) \to 0$ as $v \to \infty$. That is, $|w(u)|$ is bounded for large u, and we deduce as above that the P-probability of such w's is zero. □

Example 3.16. An example of a process $\{W(t)\}_{t\in\mathbb{R}}$ satisfying the conditions around (3.178) can be constructed as follows. Let $\{B_{\widetilde{H}}(t)\}$ be fractional Brownian motion (FBM) with self-similarity parameter $\widetilde{H}$ (see Section 1.1). FBM is not stationary but has stationary increments instead. In particular, the increment process of FBM defined as

$$W(t) = B_{\widetilde{H}}(t) - B_{\widetilde{H}}(t-1), \quad t \in \mathbb{R}$$

is stationary (see also Pipiras and Taqqu [42], Remark 2.5.3). Since FBM has all its moments finite (being Gaussian), is $\widetilde{H}$-self-similar, and has stationary increments, we get that

$$E|W(t)-W(s)|^{\alpha} \le C_{\alpha}E|B_{\widetilde{H}}(t)-B_{\widetilde{H}}(s)|^{\alpha} + C_{\alpha}E|B_{\widetilde{H}}(t-1)-B_{\widetilde{H}}(s-1)|^{\alpha}$$

$$= 2C_{\alpha}E|B_{\widetilde{H}}(t)-B_{\widetilde{H}}(s)|^{\alpha} = 2C_{\alpha}E|B_{\widetilde{H}}(t-s)|^{\alpha} = C|t-s|^{\widetilde{H}\alpha}.$$

That is, the condition (3.178) is satisfied with $p = \widetilde{H}\alpha$. By the Kolmogorov continuity criterion, FBM can be taken to have continuous sample paths (see, e.g., Pipiras and Taqqu [42], Proposition 7.1.2), and hence so the process $\{W(t)\}_{t\in\mathbb{R}}$. Since $W(t) = B_{\widetilde{H}}(t) - B_{\widetilde{H}}(t-1)$ has the same Gaussian distribution for each fixed t, it follows that $P(|W(t)| < c) < 1$ for all $c > 0$. Moreover, the stationary process $\{W(t)\}_{t\in\mathbb{R}}$ is ergodic since its autocovariance function $\gamma(h) = EW(t+h)W(t)$ can be checked to decay to 0 as lag $h \to \infty$ (see, e.g., Lindgren [30], Theorem 6.6). We thus conclude that the process $\{W(t)\}_{t\in\mathbb{R}}$ satisfies all the conditions around (3.178).

Appendix A
Historical Notes

We provide here a number of historical and other notes on the material of this work.

Section 1.3: A comprehensive treatment of stable processes and their integral representations can be found in Samorodnitsky and Taqqu [56]. Univariate stable distributions is the focus of a fascinating monograph by Uchaikin and Zolotarev [63], as well as of the classic monograph by Zolotarev [69]. For a more recent treatment of stable distributions and processes, see Rachev and Mittnik [43] for applications in finance, and Nolan [32].

We assumed in Section 1.3 and throughout the work that stable processes are $S\alpha S$. In the study of stationary stable processes and their connections to flows, *skewness* was included with "minor" modifications in Rosiński [45] and Kolodyński and Rosiński [24]. Extensions to stationary stable *fields*, that is, stable processes $X_\alpha(t)$ indexed by $t \in \mathbb{R}^d$ or $t \in \mathbb{Z}^d$, are not immediate and were considered in Rosiński [47] and Roy and Samorodnitsky [49]. Similar developments concerning skewness and random fields are expected but not available yet in the context of self-similar mixed moving averages.

Section 2.1: The definition (2.3) of minimal integral representations was given by Hardin [19, 20]. Rosiński [48] reexamined minimal integral representations in greater depth and provided alternative formulations, including those in (2.4)–(2.5) and Remark 2.1. Pipiras [34] showed that a minimal integral representation can be achieved naturally from a nonminimal one by removing a so-called nonminimal set from the space underlying the nonminimal integral representation.

Despite significant progress in the theory, determining whether a given integral representation is minimal, may not be immediate. See, for example, Examples 2.2 and 3.3 concerning stationary and stationary increments moving averages.

Section 2.2: Theorem 2.1, characterizing linear isometries of spaces L^α, appears in Banach [4] and Lamperti [29]. It is a well-known result in real analysis, for example, part of the classic by Royden [50], Chapter 15. For a more advanced and

V. Pipiras, M.S. Taqqu, *Stable Non-Gaussian Self-Similar Processes with Stationary Increments*, SpringerBriefs in Probability and Mathematical Statistics,
DOI 10.1007/978-3-319-62331-3

comprehensive treatment, including Rudin's theorem (see (2.36)–(2.37)), see Fleming and Jamison [14], Section 3. Section 3.5, in particular, has interesting notes and remarks.

The statement and proof of Proposition 2.2, relating minimal integral representations and isometries, are taken from Rosiński [48].

Section 2.3: As indicated in the introduction of Section 1.1, the connection between stationary stable processes and nonsingular flows outlined in Section 2.3.1 is due to Rosiński [46]. Nonsingular flows, cocycles, and related notions (Hopf decomposition, special flows) are part of ergodic theory (Cornfeld et al. [8], Krengel [26], Zimmer [68]), and, especially, infinite[1] ergodic theory (Aaronson [1], Glasner [15]).

1- and 2-semi-additive functionals arise specifically when relating self-similar mixed moving averages and flows. The terms are coined by analogy to additive functionals $f_h(s)$ (for a flow $\phi_h(s)$) satisfying

$$f_{h_1+h_2}(s) = f_{h_1}(s) + f_{h_2}(\phi_{h_1}(s)), \quad s \in S,\ h_1, h_2 \in \mathbb{R}$$

(Kubo [27, 28]). See Pipiras and Taqqu [41].

Having established a connection between stable processes and flows, it is natural to ask how ergodic or other properties of flows affect those of the corresponding stable processes. For stationary stable processes, this is studied in Samorodnitsky [52, 53].

Section 3.1: The name mixed moving averages is adopted from the stationary case, where mixed moving averages are represented as

$$\int_X \int_{\mathbb{R}} G(x, t+u) M(dx, du).$$

These processes were studied in Surgailis et al. [60], and are stationary stable processes generated by dissipative flows in the decomposition established by Rosiński [46].

Using a suitable transformation, Surgailis et al. [61] associated stable processes with stationary increments with stable stationary processes, and then decomposed them based on the decomposition of stationary processes of Rosiński [46]. In this approach, mixed moving averages with stationary increments are associated with stationary mixed moving averages and can be represented by (3.1). From the perspective of these structural results, it should be noted that mixed moving averages make just a single, even if large, class of stable processes with stationary increments (namely, those generated by dissipative flows).

Section 3.2: Sections 3.2.1 and 3.2.2 are based on Pipiras and Taqqu [35], Section 3.2.3 on Pipiras and Taqqu [36, 37], Section 3.2.5 on Pipiras and Taqqu [36, 40], and Section 3.2.6 on Pipiras and Taqqu [39].

[1] "Infinite" refers here to the fact that the measure on the space underlying a flow is not finite. Another term used is "nonsingular ergodic theory."

Several notable differences between Section 3.2 and these references are the following. Definition 3.2 found in [35, 36] does not stipulate for $j_c(x)$ to be a 2-semi-additive functional. By making this assumption, the current definition thus provides a description of the structure of $j_c(x)$ as a function in (c,x). This is useful, for example, in establishing canonical representations of self-similar mixed moving averages (see the proof of Theorem 3.11). To account for the change in the definition, we had to modify the proofs of the fundamental Theorem 3.1 and several other results.

Proposition 3.2 on the existence of minimal mixed moving average representations was proved in [35] under the assumption $\alpha \in (1,2)$. This assumption was then made throughout [35, 36]. The proposition was extended to the entire range $\alpha \in (0,2)$ in Pipiras [34], making it possible to assume this range throughout Section 3.2. Finally, we also note that fixed fractional stable motions (FFSMs) were called mixed linear fractional stable motions (mixed LFSMs) in [36, 39, 40]. While the latter term was motivated by the canonical representation in Theorem 3.11, the term FFSM used in Section 3.2 is in line with the nature of the underlying (that is, fixed) flow, and the analogous terms cLFSM for cyclic fractional stable motion and PFSM for periodic fractional stable motion.

Other notes: Natural connections to nonsingular flows are available not only for stable processes but also for the so-called max-stable processes. If stable distributions arise as limits of normalized partial sums, max-stable distributions arise as limits of normalized partial maxima. Max-stable processes thus play an important role in extreme value theory (Coles [7], de Haan and Ferreira [9]).

Minimal representations, max-linear isometries, and glimpses of flows in the context of max-stable processes date back at least to de Haan and Pickands [10]. A flow-based decomposition of stationary max-stable processes, parallel to that of Rosiński [46] for stationary stable processes, was established by Wang and Stoev [67]. Parallels between max-stable and stable frameworks are studied in Kabluchko [21] and Wang and Stoev [66].

Appendix B
Standard Lebesgue Spaces and Projections

A measure space $(S,\mathscr{S},\nu)$ is called a *standard Lebesgue space* when $(S,\mathscr{S})$ is a *standard Borel space*, equipped with a σ-finite measure μ. In a standard Borel space, S can be thought as a Borel subset of a Polish space, and the σ-field $\mathscr{S}$ is the σ-field of Borel sets $\mathscr{B}(S)$ defined as $\mathscr{B}(S)=\sigma\{A:A\subset S \text{ is open}\}$. An example of a standard Lebesgue space is the Euclidean space $\mathbb{R}^n$, with a measure consisting of Lebesgue measure and discrete point masses.

Standard Lebesgue spaces (or standard Borel spaces) are convenient to work with, have nice properties and are widely used in ergodic theory (Walters [65], Petersen [33]) and in other areas of mathematics (Mackey [31], Arveson [3], Zimmer [68]).

In this book, we used a number of times the notion of projection. If S_1 and S_2 are two spaces, the projection of a set $E\in S_1\times S_2$ onto S_1 is defined as

$$\mathrm{proj}_{S_1}E=\{s_1\in S_1:\exists s_2\in S_2:(s_1,s_2)\in E\}.$$

The projection $\mathrm{proj}_{S_2}E$ is defined in a similar way. When $S_1=S_2=S$, proj_S is understood as the projection onto the first variable.

When $(S_1,\mathscr{B}(S_1))$ and $(S_2,\mathscr{B}(S_2))$ are two standard Borel spaces, and $E\in\mathscr{B}(S_1)\times\mathscr{B}(S_2)$, it is well known that $\mathrm{proj}_{S_1}E$ is not necessarily in $\mathscr{B}(S_1)$. This important fact has essentially given rise to the field of the so-called (classical) descriptive set theory. See, for example, the monographs of Kechris [23] and Srivastava [59].

One of the key notions in descriptive set theory is that of *projective classes* $\Sigma_n^1(S),\Pi_n^1(S),\Delta_n^1(S)$, $n\in\mathbb{N}$, on a Polish space S. They can be defined recursively in n as follows. For $n=1$,

$$\Sigma_1^1(S)=\{B:B=\mathrm{proj}_S A \text{ for } A\in\mathscr{B}(S^2)\}, \tag{B.1}$$

$$\Pi_1^1(S)=\{B^c:B\in\Sigma_1^1(S)\},\quad \Delta_1^1(S)=\Sigma_1^1(S)\cap\Pi_1^1(S).$$

V. Pipiras, M.S. Taqqu, *Stable Non-Gaussian Self-Similar Processes with Stationary Increments*, SpringerBriefs in Probability and Mathematical Statistics,
DOI 10.1007/978-3-319-62331-3

The elements of $\Sigma_1^1(S)$ and $\Pi_1^1(S)$ are called *analytic* and *coanalytic* sets, respectively, and $\Delta_1^1(S) = \mathscr{B}(S)$. Then, recursively in n,

$$\Sigma_{n+1}^1(S) = \{B : B = \mathrm{proj}_S A \text{ for } A \in \Pi_n^1(S^2)\},$$

$$\Pi_{n+1}^1(S) = \{B^c : B \in \Sigma_{n+1}^1(S)\}, \quad \Delta_{n+1}^1(S) = \Sigma_{n+1}^1(S) \cap \Pi_{n+1}^1(S).$$

Another, more general way is to define the class Σ_{n+1}^1 as images of sets from the projective class Π_n^1 under Borel maps (projection, as in (B.1), is one such map).

Much is known about the above projective classes. For example, in the following diagram, any class is a subset of every class to the right of it:

$$\begin{array}{ccccccc} & \Sigma_1^1(S) & & \Sigma_2^1(S)\ \ldots & & \\ \mathscr{B}(S) = \Delta_1^1(S) & & \Delta_2^1(S) & \ldots & \overline{\mathscr{B}}(S) \\ & \Pi_1^1(S) & & \Pi_2^1(S)\ \ldots & & \end{array} \tag{B.2}$$

where

$$\overline{\mathscr{B}}(S) := \overline{\mathscr{B}}_\mu(S)$$

is the completion σ-field of $\mathscr{B}(S)$ under μ. The classes $\Sigma_n^1(S)$ and $\Pi_n^1(S)$ are closed under countable unions and intersections, and $\Delta_n^1(S)$ are σ-fields, and so on. Except the proof of Lemma B.1 below, the various intermediate classes in (B.2) are not used directly in the book. But they are naturally related to projections, and as discussed here, are behind a useful theory of classes in between the commonly used spaces $\mathscr{B}(S)$ and $\overline{\mathscr{B}}(S)$.

Another important idea is that of *uniformization* and *uniformizing functions* defined in Kechris [23], or measurable sections defined in Srivastava [59], or measurable selections defined in Wagner [64]. We are not going to describe the many related results here. We shall only state next an auxiliary result which was used above (e.g., Lemma 2.3). The function h appearing below is called a measurable selection.

Lemma B.1. *Let $(S_1, \mathscr{S}_1, \nu_1)$ and $(S_2, \mathscr{S}_2, \nu_2)$ be two standard Lebesgue spaces and $(S_1 \times S_2, \mathscr{S}_1 \times \mathscr{S}_2, \nu_1 \times \nu_2)$ be their Cartesian product. Let also $A \in \mathscr{S}_1 \times \mathscr{S}_2$ be a Borel set of $S_1 \times S_2$. Then, the set*

$$\mathrm{proj}_{S_1} A := \{s_1 \in S_1 : \exists s_2 \in S_2 : (s_1, s_2) \in A\}$$

is ν_1-measurable, that is, belongs to $\overline{\mathscr{B}}_{\mu_1}(S_1)$, and there is a ν_1-measurable function $h : \mathrm{proj}_{S_1} A \mapsto A$ such that $(s_1, h(s_1)) \in A$ for all $s_1 \in \mathrm{proj}_{S_1} A$.

PROOF: The set $\mathrm{proj}_{S_1} A$ is ν_1-measurable because the map $\mathrm{proj}_{S_1}(s_1, s_2) = s_1$ is continuous and the set A can be approximated $(\nu_1 \times \nu_2)$-a.e. by rectangles whose projections are measurable. (In the case $S_1 = S_2 = S$, this also follows from the inclusions in (B.2) where all the classes, including $\Sigma_1^1(S)$, are subsets of $\overline{\mathscr{B}}(S)$.)

We will show next that there is a ν_1-measurable map $h : \mathrm{proj}_{S_1} A \mapsto A$ such that $(s_1, h(s_1)) \in A$ for $s_1 \in \mathrm{proj}_{S_1} A$. To do so, we will use Theorem 3.4.3 in Arveson [3],

p. 77, which concerns the so-called cross sections of Borel maps. Consider the map $f = \text{proj}_{S_1} : A \mapsto f(A) = \text{proj}_{S_1} A$. The image set $f(A)$, together with the induced Borel structure

$$\mathscr{F}(A) = \{f(A) \cap B : B \in \mathscr{S}_1\},$$

is a Borel space. Moreover, this Borel space is countably separated (as defined in Arveson [3], p. 69) since the underlying standard Lebesgue space $(S_1, \mathscr{S}_1, \nu_1)$ is countably separated. The Borel set A, equipped with the Borel structure

$$\mathscr{A} = \{A \cap B : B \in \mathscr{S}_1 \times \mathscr{S}_2\},$$

is also a Borel space. It is an analytic Borel space (as defined in Arveson [3], p. 71) by using Corollary in Arveson [3], p. 65, and the fact that A is a Borel set. Since $f^{-1}(f(A) \cap B) = A \cap (B \times \mathbb{R}) \in \mathscr{A}$ for all $B \in \mathscr{S}_1$, the map

$$f : (A, \mathscr{A}) \mapsto (f(A), \mathscr{F}(A))$$

is Borel. It follows from Theorem 3.4.3 in Arveson [3] that there is a ν_1-measurable map $g : f(A) \mapsto A$ such that $f(g(s_1)) = s_1$. Since f is a projection, we have that $g(s_1) = (s_1, h(s_1))$ for some ν-measurable map $h(s_1)$ and hence that there is a ν-measurable map $h(s_1)$ such that $(s_1, h(s_1)) \in A$. $\square$

Appendix C
Notation Summary

We denote by

$$X_\alpha^D,\ X_\alpha^C,\ X_\alpha^P,\ X_\alpha^F,\ X_\alpha^L,\ X_\alpha^{C\setminus P}$$

the processes generated respectively by dissipative (D), conservative (C), periodic (P), fixed (F), cyclic (L), and conservative but not periodic ($C \setminus P$) flows. These processes are abbreviated respectively

$$\text{DFSM, CFSM, PFSM, FFSM, cLFSM, (C}\setminus\text{P)FSM.}$$

They are characterized in terms of minimal representations. The dissipative (D) and conservative (C) sets associated with a flow can also be characterized through the structure of the kernel function G in (3.1). The processes

$$X_\alpha^P,\ X_\alpha^F,\ X_\alpha^L,\ X_\alpha^{C\setminus P}$$

can also be characterized through the structure of the kernel function G in (3.1), namely through respectively, the following subsets of C,

$$C_P,\ C_F,\ C_L,\ C \setminus C_P.$$

Note that the linear fractional stable motion (LFSM) defined in Example 2.5 is also a FFSM as indicated in Example 3.14.

V. Pipiras, M.S. Taqqu, *Stable Non-Gaussian Self-Similar Processes with Stationary Increments*, SpringerBriefs in Probability and Mathematical Statistics,
DOI 10.1007/978-3-319-62331-3

References

[1] J. Aaronson. *An Introduction to Infinite Ergodic Theory*, volume 50 of *Mathematical Surveys and Monographs*. American Mathematical Society, Providence, RI, 1997.
(Cited on page 116.)

[2] J. Aaronson and M. Denker. Characteristic functions of random variables attracted to 1-stable laws. *The Annals of Probability*, 26(1):399–415, 1998.
(Cited on page 4.)

[3] W. Arveson. *An Invitation to C^*-Algebras*. Springer-Verlag, New York, 1976.
(Cited on pages 119, 120, 121.)

[4] S. Banach. *Théorie des Opérations Linéaires*. Warsaw, 1932.
(Cited on pages 17, 115.)

[5] N. H. Bingham, C. M. Goldie, and J. L. Teugels. *Regular Variation*, volume 27 of *Encyclopedia of Mathematics and its Applications*. Cambridge University Press, Cambridge, 1987.
(Cited on pages 40, 46.)

[6] P. J. Brockwell and R. A. Davis. *Time Series: Theory and Methods*. Springer Series in Statistics. Springer-Verlag, New York, second edition, 1991.
(Cited on page 32.)

[7] S. Coles. *An Introduction to Statistical Modeling of Extreme Values*. Springer Series in Statistics. Springer-Verlag London Ltd., London, 2001.
(Cited on page 117.)

[8] I. P. Cornfeld, S. V. Fomin, and Ya. G. Sinai. *Ergodic Theory*. Springer-Verlag, 1982.
(Cited on page 116.)

[9] L. de Haan and A. Ferreira. *Extreme Value Theory*. Springer Series in Operations Research and Financial Engineering. Springer, New York, 2006. An introduction.
(Cited on page 117.)

V. Pipiras, M.S. Taqqu, *Stable Non-Gaussian Self-Similar Processes with Stationary Increments*, SpringerBriefs in Probability and Mathematical Statistics,
DOI 10.1007/978-3-319-62331-3

[10] L. de Haan and J. Pickands III. Stationary min-stable stochastic processes. *Probability Theory and Related Fields*, 72(4):477–492, 1986.
(Cited on page 117.)

[11] J. L. Doob. *Stochastic Processes*. Wiley Classics Library. John Wiley & Sons Inc., New York, 1953. Reprint of the 1953 original, a Wiley-Interscience Publication.
(Cited on page 49.)

[12] H. Dym and H. P. McKean. *Fourier Series and Integrals*. Academic Press, New York, 1972.
(Cited on page 18.)

[13] P. Embrechts and M. Maejima. *Selfsimilar Processes*. Princeton Series in Applied Mathematics. Princeton University Press, 2002.
(Cited on pages 1, 2.)

[14] R. J. Fleming and J. E. Jamison. *Isometries on Banach Spaces: Function Spaces*, volume 129 of *Chapman & Hall/CRC Monographs and Surveys in Pure and Applied Mathematics*. Chapman & Hall/CRC, Boca Raton, FL, 2003.
(Cited on page 116.)

[15] E. Glasner. *Ergodic Theory via Joinings*, volume 101 of *Mathematical Surveys and Monographs*. American Mathematical Society, Providence, RI, 2003.
(Cited on page 116.)

[16] B. V. Gnedenko and A. N. Kolmogorov. *Limit distributions for sums of independent random variables*. Addison-Wesley, Reading, MA, 1954.
(Cited on page 4.)

[17] K. Górska and K. A. Penson. Lévy stable two-sided distributions: Exact and explicit densities for asymmetric case. *Physical Review E*, 83:061125, 2011.
(Cited on page 3.)

[18] P. R. Halmos. *Measure Theory*. Van Nostrand, New York, 1950.
(Cited on page 40.)

[19] C. D. Hardin Jr. Isometries on subspaces of L^p. *Indiana University Mathematics Journal*, 30:449–465, 1981.
(Cited on pages 22, 23, 115.)

[20] C. D. Hardin Jr. On the spectral representation of symmetric stable processes. *Journal of Multivariate Analysis*, 12:385–401, 1982.
(Cited on pages 11, 12, 115.)

[21] Z. Kabluchko. Spectral representations of sum- and max-stable processes. *Extremes*, 12(4):401–424, 2009.
(Cited on page 117.)

[22] M. Kanter. The L^p norm of sums of translates of a function. *Transactions of the American Mathematical Society*, 79:35–47, 1973.
(Cited on page 33.)

[23] A. S. Kechris. *Classical Descriptive Set Theory*. Springer-Verlag, New York, 1995.
(Cited on pages 119, 120.)

[24] S. Kolodyński and J. Rosiński. Group self-similar stable processes in $\mathbb{R}^d$. *Journal of Theoretical Probability*, 16(4):855–876 (2004), 2003.
(Cited on page 115.)
[25] U. Krengel. Darstellungssätze für Strömungen und Halbströmungen II. *Mathematische Annalen*, 182:1–39, 1969.
(Cited on pages 37, 38.)
[26] U. Krengel. *Ergodic Theorems*. Walter de Gruyter, Berlin, 1985.
(Cited on pages 36, 116.)
[27] I. Kubo. Quasi-flows. *Nagoya Mathematical Journal*, 35:1–30, 1969.
(Cited on pages 42, 116.)
[28] I. Kubo. Quasi-flows II: Additive functionals and TQ-systems. *Nagoya Mathematical Journal*, 40:39–66, 1970.
(Cited on pages 45, 46, 48, 116.)
[29] J. Lamperti. On the isometries of certain function-spaces. *Pacific Journal of Mathematics*, 8:459–466, 1958.
(Cited on pages 17, 115.)
[30] G. Lindgren. *Stationary Stochastic Processes*. Chapman & Hall/CRC Texts in Statistical Science Series. CRC Press, Boca Raton, FL, 2013. Theory and applications.
(Cited on page 114.)
[31] G. W. Mackey. Borel structure in groups and their duals. *Transactions of the American Mathematical Society*, 85:134–165, 1957.
(Cited on pages 63, 119.)
[32] J. P. Nolan. *Stable Distributions - Models for Heavy Tailed Data*. Forthcoming, 2017.
(Cited on page 115.)
[33] K. Petersen. *Ergodic Theory*. Cambridge University Press, Cambridge, 1983.
(Cited on page 119.)
[34] V. Pipiras. Nonminimal sets, their projections and integral representations of stable processes. *Stochastic Processes and their Applications*, 117(9):1285–1302, 2007.
(Cited on pages 53, 115, 117.)
[35] V. Pipiras and M. S. Taqqu. Decomposition of self-similar stable mixed moving averages. *Probability Theory and Related Fields*, 123(3):412–452, 2002.
(Cited on pages 45, 51, 53, 78, 109, 116, 117.)
[36] V. Pipiras and M. S. Taqqu. The structure of self-similar stable mixed moving averages. *The Annals of Probability*, 30(2):898–932, 2002.
(Cited on pages 87, 94, 95, 96, 108, 109, 116, 117.)
[37] V. Pipiras and M. S. Taqqu. Dilated fractional stable motions. *Journal of Theoretical Probability*, 17(1):51–84, 2004.
(Cited on pages 78, 81, 116.)
[38] V. Pipiras and M. S. Taqqu. Stable stationary processes related to cyclic flows. *The Annals of Probability*, 32(3A):2222–2260, 2004.
(Cited on page 41.)

[39] V. Pipiras and M. S. Taqqu. Integral representations of periodic and cyclic fractional stable motions. *Electronic Journal of Probability*, 12:no. 7, 181–206, 2007.
(Cited on pages 101, 104, 105, 116, 117.)

[40] V. Pipiras and M. S. Taqqu. Identification of periodic and cyclic fractional stable motions. *Annales de l'Institut Henri Poincaré Probabilités et Statistiques*, 44(4):612–637, 2008.
(Cited on pages 91, 94, 95, 109, 111, 116, 117.)

[41] V. Pipiras and M. S. Taqqu. Semi-additive functionals and cocycles in the context of self-similarity. *Discussiones Mathematicae. Probability and Statistics*, 30(2):149–177, 2010.
(Cited on pages 48, 64, 116.)

[42] V. Pipiras and M. S. Taqqu. *Long-Range Dependence and Self-Similarity*. Cambridge University Press, 2017.
(Cited on pages 1, 2, 4, 31, 50, 114.)

[43] S. T. Rachev and S. Mittnik. *Stable Paretian Models in Finance*. Wiley, New York, 2000.
(Cited on page 115.)

[44] J. Rosiński. On uniqueness of the spectral representation of stable processes. *Journal of Theoretical Probability*, 7(3):615–634, 1994.
(Cited on pages 20, 21, 24, 28, 29, 30.)

[45] J. Rosiński. Uniqueness of spectral representations of skewed stable processes and stationarity. In H. Kunita and H.-H. Kuo, editors, *Stochastic Analysis On Infinite Dimensional Spaces.*, pages 264–273. Proceedings of the U.S.-Japan Bilateral Seminar, 1994.
(Cited on pages 5, 115.)

[46] J. Rosiński. On the structure of stationary stable processes. *The Annals of Probability*, 23:1163–1187, 1995.
(Cited on pages 2, 36, 37, 38, 53, 63, 116, 117.)

[47] J. Rosiński. Decomposition of stationary α-stable random fields. *The Annals of Probability*, 28(4):1797–1813, 2000.
(Cited on page 115.)

[48] J. Rosiński. Minimal integral representations of stable processes. *Probability and Mathematical Statistics*, 26(1):121–142, 2006.
(Cited on pages 11, 15, 16, 24, 115, 116.)

[49] P. Roy and G. Samorodnitsky. Stationary symmetric α-stable discrete parameter random fields. *Journal of Theoretical Probability*, 21(1):212–233, 2008.
(Cited on page 115.)

[50] H. L. Royden. *Real Analysis*. Macmillan, third edition, 1988.
(Cited on pages 19, 115.)

[51] W. Rudin. L^p-isometries and equimeasurability. *Indiana University Mathematics Journal*, 25(3):215–228, 1976.
(Cited on page 22.)

[52] G. Samorodnitsky. Extreme value theory, ergodic theory and the boundary between short memory and long memory for stationary stable processes. *The Annals of Probability*, 32(2):1438–1468, 2004.
(Cited on page 116.)

[53] G. Samorodnitsky. Null flows, positive flows and the structure of stationary symmetric stable processes. *The Annals of Probability*, 33(5):1782–1803, 2005.
(Cited on page 116.)

[54] G. Samorodnitsky. *Stochastic Processes and Long Range Dependence*. Springer Series in Operations Research and Financial Engineering, Springer, 2016.
(Cited on page 2.)

[55] G. Samorodnitsky and M. S. Taqqu. $1/\alpha$-self-similar processes with stationary increments.. *Journal of Multivariate Analysis*, 35:308–313, 1990.
(Cited on pages 80, 84.)

[56] G. Samorodnitsky and M. S. Taqqu. *Stable Non-Gaussian Random Processes*. Stochastic Modeling. Chapman & Hall, New York, 1994. Stochastic models with infinite variance.
(Cited on pages 2, 3, 4, 5, 6, 7, 8, 9, 12, 31, 33, 72, 79, 85, 115.)

[57] K.-i. Sato. *Lévy Processes and Infinitely Divisible Distributions*, volume 68 of *Cambridge Studies in Advanced Mathematics*. Cambridge University Press, Cambridge, 2013. Translated from the 1990 Japanese original, Revised edition of the 1999 English translation.
(Cited on page 4.)

[58] R. Sikorski. *Boolean Algebras*. Third edition. Ergebnisse der Mathematik und ihrer Grenzgebiete, Band 25. Springer-Verlag New York Inc., New York, 1969.
(Cited on page 20.)

[59] S. M. Srivastava. *A Course on Borel Sets*. Springer-Verlag, New York, 1998.
(Cited on pages 119, 120.)

[60] D. Surgailis, J. Rosiński, V. Mandrekar, and S. Cambanis. Stable generalized moving averages. *Probability Theory and Related Fields*, 97:543–558, 1993.
(Cited on page 116.)

[61] D. Surgailis, J. Rosiński, V. Mandrekar, and S. Cambanis. On the mixing structure of stationary increment and self-similar $S\alpha S$ processes. Preprint, 1998.
(Cited on pages 80, 116.)

[62] S. Takenaka. Integral-geometric construction of self-similar stable processes. *Nagoya Mathematical Journal*, 123:1–12, 1991.
(Cited on page 79.)

[63] V. V. Uchaikin and V. M. Zolotarev. *Chance and Stability*. Modern Probability and Statistics. VSP, Utrecht, 1999. Stable distributions and their applications, With a foreword by V. Yu. Korolev and Zolotarev.
(Cited on pages 3, 4, 115.)

[64] D. H. Wagner. Survey of measurable selection theorems. *SIAM Journal of Control and Optimization*, 15(5):859–903, 1977.
(Cited on page 120.)

[65] P. Walters. *An Introduction to Ergodic Theory*. Springer-Verlag, New York, 1982.
(Cited on page 119.)

[66] Y. Wang and S. A. Stoev. On the association of sum- and max-stable processes. *Statistics & Probability Letters*, 80(5–6):480–488, 2010.
(Cited on page 117.)

[67] Y. Wang and S. A. Stoev. On the structure and representations of max-stable processes. *Advances in Applied Probability*, 42(3):855–877, 2010.
(Cited on page 117.)

[68] R. J. Zimmer. *Ergodic Theory and Semisimple Groups*. Birkhäuser, Boston, 1984.
(Cited on pages 45, 116, 119.)

[69] V. M. Zolotarev. *One-dimensional Stable Distributions*, volume 65 of "Translations of mathematical monographs". American Mathematical Society, 1986. Translation from the original 1983 Russian edition.
(Cited on pages 3, 4, 115.)

Author Index

V. Pipiras, M.S. Taqqu, *Stable Non-Gaussian Self-Similar Processes with Stationary Increments*, SpringerBriefs in Probability and Mathematical Statistics,
DOI 10.1007/978-3-319-62331-3

Subject Index

V. Pipiras, M.S. Taqqu, *Stable Non-Gaussian Self-Similar Processes with Stationary Increments*, SpringerBriefs in Probability and Mathematical Statistics,
DOI 10.1007/978-3-319-62331-3

Printed in the United States
By Bookmasters